RESTRUCTURING FEDERAL CLIMATE RESEARCH TO MEET THE CHALLENGES OF
CLIMATE CHANGE

Committee on Strategic Advice on the
U.S. Climate Change Science Program

Division on Earth and Life Studies

Division of Behavioral and Social Sciences and Education

NATIONAL RESEARCH COUNCIL
OF THE NATIONAL ACADEMIES

THE NATIONAL ACADEMIES PRESS
Washington, D.C.
www.nap.edu

THE NATIONAL ACADEMIES PRESS • 500 Fifth Street, N.W. • Washington, DC 20001

NOTICE: The project that is the subject of this report was approved by the Governing Board of the National Research Council, whose members are drawn from the councils of the National Academy of Sciences, the National Academy of Engineering, and the Institute of Medicine. The members of the committee responsible for the report were chosen for their special competences and with regard for appropriate balance.

This study was supported by the National Aeronautics and Space Administration under Award No. NNH07CC79B. Any opinions, findings, conclusions, or recommendations expressed in this publication are those of the author(s) and do not necessarily reflect the view of the organizations or agencies that provided support for this project.

International Standard Book Number-13: 978-0-309-13173-5
International Standard Book Number-10: 0-309-13173-1

Library of Congress Control Number: 2009923757

Additional copies of this report are available from the National Academies Press, 500 Fifth Street, N.W., Lockbox 285, Washington, DC 20055; (800) 624-6242 or (202) 334-3313 (in the Washington metropolitan area); Internet http://www.nap.edu.

Cover: Designed by Van Nguyen.

THE NATIONAL ACADEMIES
Advisers to the Nation on Science, Engineering, and Medicine

The **National Academy of Sciences** is a private, nonprofit, self-perpetuating society of distinguished scholars engaged in scientific and engineering research, dedicated to the furtherance of science and technology and to their use for the general welfare. Upon the authority of the charter granted to it by the Congress in 1863, the Academy has a mandate that requires it to advise the federal government on scientific and technical matters. Dr. Ralph J. Cicerone is president of the National Academy of Sciences.

The **National Academy of Engineering** was established in 1964, under the charter of the National Academy of Sciences, as a parallel organization of outstanding engineers. It is autonomous in its administration and in the selection of its members, sharing with the National Academy of Sciences the responsibility for advising the federal government. The National Academy of Engineering also sponsors engineering programs aimed at meeting national needs, encourages education and research, and recognizes the superior achievements of engineers. Dr. Charles M. Vest is president of the National Academy of Engineering.

The **Institute of Medicine** was established in 1970 by the National Academy of Sciences to secure the services of eminent members of appropriate professions in the examination of policy matters pertaining to the health of the public. The Institute acts under the responsibility given to the National Academy of Sciences by its congressional charter to be an adviser to the federal government and, upon its own initiative, to identify issues of medical care, research, and education. Dr. Harvey V. Fineberg is president of the Institute of Medicine.

The **National Research Council** was organized by the National Academy of Sciences in 1916 to associate the broad community of science and technology with the Academy's purposes of furthering knowledge and advising the federal government. Functioning in accordance with general policies determined by the Academy, the Council has become the principal operating agency of both the National Academy of Sciences and the National Academy of Engineering in providing services to the government, the public, and the scientific and engineering communities. The Council is administered jointly by both Academies and the Institute of Medicine. Dr. Ralph J. Cicerone and Dr. Charles M. Vest are chair and vice chair, respectively, of the National Research Council.

www.national-academies.org

Preface

The U.S. Climate Change Science Program (CCSP) is developing a new strategic plan to replace the one that has guided federal research since 2003. The new strategic plan is expected to be released early in the next administration. There is thus an opportunity to step back, examine what has been learned, and chart a new course for the future. The National Research Council's Committee on Strategic Advice on the U.S. Climate Change Science Program was established to evaluate progress of the CCSP and to identify future priorities. Its first report, *Evaluating Progress of the U.S. Climate Change Science Program: Methods and Preliminary Results* (NRC, 2007c), drew the following conclusions about the progress of the CCSP:

- Discovery science and understanding of the climate system are proceeding well, but use of that knowledge to support decision making and to manage risks and opportunities of climate change is proceeding slowly
- Progress in understanding and predicting climate change has improved more at global, continental, and ocean basin scales than at regional and local scales

• Our understanding of the impact of climate changes on human well-being and vulnerabilities is much less developed than our understanding of the natural climate system

• Science quality observation systems have fueled advances in climate change science and applications, but many existing and planned observing systems have been cancelled, delayed, or degraded, which threatens future progress

• Progress in communicating CCSP results and engaging stakeholders is inadequate

• The separation of leadership and budget authority presents a serious obstacle to progress in the CCSP

This is the second report and it identifies priorities for addressing these issues and for meeting new scientific and societal needs. To gather input and discuss the issues, the committee held five meetings and two major workshops. Most of the meetings were focused on particular issues, including priorities for CCSP components and for the program as a whole, and communicating scientific understanding for management and policy making. The first workshop focused on stakeholders and applied research, regional modeling, and data needed to support adaptation and mitigation in various sectors, climate policy, and national assessments (see Appendix F for the agenda and list of participants). The second workshop focused on basic natural and social science research, ways to balance competing priorities, and ways to make an interagency coordinated program work (Appendix F).

The committee also solicited essays from colleagues. Of particular note are the comprehensive summaries of research priorities in the natural sciences and the human dimensions prepared by the chair and staff of the Committee on the Human Dimensions of Global Change and the Climate Research Committee (Appendixes D and E). The committee extends its thanks to those committees and especially to the chairs (Thomas Wilbanks and Antonio Busalacchi) and staff (Ian Kraucunas and Paul Stern). Other colleagues who contributed substantial material or helped the committee sort through ideas include Dan Brown, Michael Hanemann, David Skole, and Kirk Smith. The committee greatly appreciates their contributions.

The committee also thanks the many other individuals who gave presentations, led working group discussions, or provided other input to the committee: Rick Anthes, Peter Backlund, Roberta Balstad, Bruce Barkstrom, Jonathan Black, William Brennan, Dixon Butler, L. Greg Carbone, DeWayne Cecil, Javade Chaudhri, Eileen Claussen, Andrew Comrie, Kevin Cook, Lisa Dilling, George Eads, William Easterling, Jae Edmonds, Jack Fellows, Guido Franco, Sharon Hays, Issac Held, Anthony Janetos, Timothy Killeen, Chet Koblinsky, Martha Krebs, Kent Laborde, Dennis Lettenmaier, Ruby Leung, Roger Lukas, Alexander MacDonald, Linda Mearns, Susanne Moser, Jon Padgham, Adam Phillips, Roger Pielke Jr., Andrew Revkin, Sherwood Rowland, Jason Samenow, David Schimel, Stephen Schneider, Peter Schultz, Susan Solomon, Michael Stephens, Susan Tierney, Kevin Trenberth, Compton Tucker, Robert Waterman, Anne Watkins, and Julie Winkler. Finally, the committee chair, vice chair, and the entire committee express their deep gratitude to Anne Linn, the study director, and the other NRC staff for their outstanding work in organizing the workshops and preparing the report and guiding it through the review and publication process.

V. Ramanathan, Chair
C. Justice, Vice Chair

Acknowledgments

This report has been reviewed in draft form by individuals chosen for their diverse perspectives and technical expertise, in accordance with procedures approved by the NRC's Report Review Committee. The purpose of this independent review is to provide candid and critical comments that will assist the institution in making its published report as sound as possible and to ensure that the report meets institutional standards for objectivity, evidence, and responsiveness to the study charge. The review comments and draft manuscript remain confidential to protect the integrity of the deliberative process. We wish to thank the following individuals for their participation in the review of this report:

Richard A. Anthes, University Corporation for Atmospheric Research, Boulder, Colorado
Robert H. Austin, Princeton University, New Jersey
Edward A. Boyle, Massachusetts Institute of Technology, Cambridge
F. Stuart Chapin, University of Alaska, Fairbanks
Grant Davis, Sonoma County Water Agency, Santa Rosa, California
Mark Fahnestock, University of New Hampshire, Durham
Margaret S. Leinen, Climos, Inc., Alexandria, Virginia

Linda O. Mearns, National Center for Atmospheric Research, Boulder, Colorado

M. Granger Morgan, Carnegie Mellon University, Pittsburgh, Pennsylvania

Susanne C. Moser, Susanne Moser Research & Consulting, Santa Cruz, California

William D. Nordhaus, Yale University, New Haven, Connecticut

Jonathan A. Patz, University of Wisconsin, Madison

Although the reviewers listed above have provided many constructive comments and suggestions, they were not asked to endorse the conclusions or recommendations nor did they see the final draft of the report before its release. The review of this report was overseen by Kenneth H. Brink, Woods Hole Oceanographic Institution, and Marcia K. McNutt, Monterey Bay Aquarium Research Institute. Appointed by the National Research Council, they were responsible for making certain that an independent examination of this report was carried out in accordance with institutional procedures and that all review comments were carefully considered. Responsibility for the final content of this report rests entirely with the authoring committee and the institution.

Contents

Summary

Climate change is one of the most important global environmental problems facing the world today. Evidence of a changing climate is all around us, from rising sea level to retreating mountain glaciers, melting Arctic sea ice, lengthening growing seasons, shifting animal migration patterns, and other changes. Such changes are already having adverse impacts on people's well-being, as climate change amplifies the effects of other environmental and socioeconomic changes and problems and produces new effects of its own. The long-lived greenhouse gases already in the atmosphere guarantee that warming will continue, even if emissions are drastically cut today. But emissions continue to grow as population and consumption increases. The rising demand for energy, transportation, and food are projected to further raise emissions of greenhouse gases.

Based on these trends, the Intergovernmental Panel on Climate Change has predicted that the warming during this century will be in the range of 1.5°C to 4.5°C, and likely at or close to the upper level if aggressive actions are not taken to mitigate CO_2 emissions. At a minimum, the coming decades will continue warming beyond what societies have experienced in the past, likely causing disruptive shifts in supplies of freshwater and food, increased degradation of land and ocean ecosystems, and new threats to public health, the economy, and national security. If the projected warming is abrupt, as has happened at times earlier in the planet's history, it could pose formidable challenges for adaptation measures. In the worst case, warming may trigger tipping points—thresholds for irreversible changes in the way Earth's climate operates and how human and ecological systems respond.

Given this scenario, it is likely going to be a Herculean task to limit climate change to 2°C of warming from preindustrial levels as desired by many governments. The 1997 Kyoto Protocol was an important initial step toward attempting to manage greenhouse gas emissions at the international level. At the national level, nearly 80 percent of U.S. states have adopted or are preparing climate action plans, some of which include mitigation measures such as cap and trade programs. However, many policy decisions on mitigation and adaptation are being made without the scientific support that could help shape better outcomes. Robust and effective responses to climate change demand a vastly improved body of scientific knowledge, including observations and better understanding and predictions of the changing climate system, the human drivers of climate change, the response of the climate system to these drivers, and the response of society to climate changes.

The research, observations, and modeling needed to develop the knowledge foundation for understanding and responding to climate change at the federal level is the responsibility of the U.S. Climate Change Science Program (CCSP). At the request of Dr. James Mahoney, then director of the CCSP, the National Research Council established a committee to carry out two tasks over a 3-year period. The report on the committee's first task, *Evaluating Progress of the U.S. Climate Change Science Program: Methods and Preliminary Results*, was published in 2007 (NRC, 2007c). The second task—future priorities for the program—is the subject of this report:

> *Task 2. The committee will examine the program elements described in the Climate Change Science Program strategic plan and identify priorities to guide the future evolution of the program in the context of established scientific and societal objectives. These priorities may include adjustments to the balance of science and applications, shifts in emphasis given to the various scientific themes, and identification of program*

elements not supported in the past. A report identifying these future priorities will be prepared. The recommendations will specify which priorities could likely be addressed through an evolution of existing activities or reprogramming, and which would likely require new resources or partnerships.

This report lays out a framework for generating the knowledge to understand and respond to climate change, and identifies priorities for a restructured climate change research program.

A NEW FRAMEWORK TO MEET THE CHALLENGES OF CLIMATE CHANGE

Dealing with climate change will be one of the biggest challenges of the next century. The future (post-CCSP) climate change research program will play a key role by building knowledge, through sound science and incontrovertible observations, that informs decision making. However, meeting the needs of decision makers requires a transformational change in how climate change research is organized and incorporated into public policy in the United States.

The traditional approach of organizing climate change research by scientific disciplines (e.g., atmospheric chemistry) or biophysical processes (e.g., carbon cycle) has led to significant advances in our understanding of the climate system and the creation of a robust observations and modeling infrastructure. However, the paucity of social science research and the separation of natural and social science research within the CCSP, as well as the insufficient engagement of policy makers, resource managers, and other stakeholders in the program are hindering our ability to address the problems that face society. Solving these problems requires research on the end-to-end climate change problem, from understanding causes and processes to supporting actions needed to cope with the impending societal problems of climate change. Examples of societally-important issues where an end-to-end approach is needed include (1) extreme weather and climate events and disasters, (2) sea level rise and melting ice, (3) freshwater availability, (4) agriculture and food security, (5) managing ecosystems, (6) human health, and (7) impacts on the economy of the United States. Addressing these issues requires the integration of disciplinary and multidisciplinary research, natural and social science, and basic research and practical applications.

The committee recommends that the program be restructured so that the existing CCSP research elements (e.g., atmospheric composition) and crosscutting themes (e.g., modeling, observations) contribute directly, although not exclusively, to critical scientific-societal issues such as freshwater availability, extreme weather, and sea level rise. The goal should be to evolve the program in a way that maintains the current strengths of understanding and predicting climate change, while building the capability to achieve the CCSP's vision of "a nation and the global community empowered with the science-based knowledge to manage the risks and opportunities of change in the climate and related environmental systems." Such a restructuring around scientific-societal issues is required to help the program become more cross disciplinary, more fully embrace the human dimensions component, and encourage an end-to-end approach (from basic science to decision support). It should also help the participating agencies better integrate their programs.

TOP PRIORITIES

The committee's top six priorities, cast as actions for the restructured climate change research program, are listed below. They are presented as a logical flow of actions, although work can begin on all of them simultaneously. All are necessary to establish a coherent program that provides the scientific basis for understanding climate change and developing informed responses.

Reorganize the program around integrated scientific-societal issues to facilitate crosscutting research focused on understanding the interactions among the climate, human, and environmental systems and on supporting societal responses to climate change.

Societal concerns about climate focus on changes that are visible now (e.g., melting ice) and the impacts of these changes (e.g., cost of long-term drought on agricultural production or the availability of freshwater). Addressing such societal concerns requires a strong underpinning of observations and models, strengthened research across the board—particularly in the human dimensions of global change and in user-driven (applied) research that supports decision making—and increased involvement of stakeholders (e.g., federal, state, and local government agencies; the private sector; environmental organizations).

Establish a U.S. climate observing system, defined as including physical, biological, and social observations, to ensure that data needed to address climate change are collected or continued.

The satellite and ground observing systems that fueled our current understanding of the climate system are in decline, even as demand for data capable of detecting climate variability and change is growing. Sustained, multidecadal observations of physical, biological, and social processes are required to document, understand, and predict climate change at the temporal and spatial scales relevant to federal, state, and local-level stakeholders and partner international programs. Consequently, the current satellite, land, ocean, and atmosphere observations of the climate system need to be continued and augmented. New observations are also needed—including those to support human dimensions research for developing and assessing mitigation and adaptation strategies—and existing human-social data need to be better organized and coordinated with physical climate observations to enable integrated social-natural systems research.

Climate-related observations are made by different federal and state government agencies, often to meet their own monitoring requirements. Although an interagency working group is developing a list of high-priority observations, the CCSP has not yet adopted one. But even with a list of observation priorities, the CCSP lacks the authority to direct individual agencies to collect, modify, or maintain them. Rather than relying on the voluntary contributions of participating agencies, a more strategic approach to data collection, distribution, and maintenance is needed—one that requires agencies to work together to design and implement a climate observing system.

Develop the science base and infrastructure to support a new generation of coupled Earth system models to improve attribution and prediction of high-impact regional weather and climate, to initialize seasonal-to-decadal climate forecasting, and to provide predictions of impacts affecting adaptive capacities and vulnerabilities of environmental and human systems.

Further climate change is inevitable, even if humans significantly reduce greenhouse gas emissions. It is therefore essential not only to have the capacity to explain what is happening to climate and why (attribution), but also to improve predictions of weather and climate variability at the spatial and temporal scales appropriate to assess the impacts of climate change. Both will require improved infrastructure and techniques in modeling the coupled human–land–ocean–atmosphere system, supported by sustained climate observations. The latter are necessary to further develop and constrain the models and to start model predictions from the most accurate observed state possible (initialization). Tools are also needed to translate the data and model output into information more usable by stakeholders. Improved predictions of regional climate will also require more unified modeling frameworks that provide for the hierarchical treatment of climate and forecast phenomena across a wide range of space and time scales, and for the routine production of decadal regional climate predictions at scales down to a few kilometers. New computing configurations will be needed to deal with the computational and data storage demands arising from decadal simulations at high resolution with high output frequency.

Strengthen research on adaptation, mitigation, and vulnerability.

Adaptation and mitigation strategies depend on an understanding of climate trends (including improved predictions of future climate change and extreme events), of differential vulnerabilities and adaptive capacities to climate impacts (including sensitivities and thresholds and barriers to adaptation), of economic costs and dynamics, and of human behaviors, policy preferences, and choices; and on assumptions about the future availability of technologies for reducing emissions (including cobenefits and unintended consequences of mitigation). Yet the underlying human dimensions research needed to understand and develop sound adaptation strategies is a major gap in the CCSP. Although adaptation, mitigation, and vulnerability research would be needed for all the societal issues in the proposed new research framework, an additional focused research effort would help speed results. A critical step in the process is for agencies with appropriate expertise to increase funding and take a leadership role in supporting, managing, and directing this research.

Initiate a national assessment process with broad stakeholder participation to determine the risks and costs of climate change impacts on the United States and to evaluate options for responding.

A comprehensive national assessment with periodic reporting provides a mechanism to build communication with stakeholder groups and to identify evolving science and societal needs and priorities. A useful assessment does not merely summarize published studies, but has the ability to undertake targeted research to produce new insights, observations, models, and decision support services. Results of the assessment could be used to help determine priorities for federal research on impacts, mitigation, and adaptation; provide a focus for integrated science-policy assessments and enhanced regional modeling and predictions; and build human and institutional capacity to support decision making. Although the CCSP is mandated to carry out a national assessment every 4 years, the last one to involve a broad range of stakeholders was conducted a decade ago. From 2006 to 2008, the CCSP published 21 synthesis and assessment reports on a range of topics and an overarching synthesis. Although useful, the collection does not add up to a comprehensive national assessment. A new assessment will require strong political and scientific leadership, adequate resources, a careful planning process, and engagement of stakeholders at all stages of the process.

Coordinate federal efforts to provide climate services (scientific information, tools, and forecasts) routinely to decision makers.

Demand is growing for credible, understandable, and useful information for responding to climate change. A comprehensive approach to supporting decisions on climate change includes two-way communication with users to determine their information needs, provision of climate services, and research to support the services. Although a few pilot efforts are providing selected climate services, a national program to monitor climate trends and issue predictions to support decision makers at multiple levels and in the various sectors has yet to be established. A national climate service should probably reside outside of the future climate change research program for a variety of reasons, including the potential to overwhelm the research program with myriad demands for specialized services. Regardless of where the service is established, the restructured climate change research program would have to be involved in the research and development of experimental products (e.g., regional predictions), tools (e.g., models), and outreach services needed to support stakeholders. The climate service could then use the tools to create products operationally. Maintaining strong links to the research program would also help the climate service take advantage of new capabilities.

PROGRAMMATIC AND BUDGET IMPLICATIONS

Implementing the above priorities will require good leaders at all levels with the authority to direct budgets and/or research efforts. Of particular importance are strong, charismatic, scientifically respected leaders for the overall program (to advocate for program goals) and for the human dimensions (to help steer the program toward a more comprehensive view of the climate–human–environmental system). A successful program also requires strong support from the White House, particularly from the Office of Science and Technology Policy to facilitate coordination with related federal programs, and from the Office of Management and Budget to secure funding for key priorities. The recent appointments of a climate czar and agency leaders interested in the climate-energy nexus create an opportunity for carrying out the transformational climate-change research envisioned by the committee as well as for strengthening coordination of climate change science and technology across the federal government.

CCSP funding has been declining since its peak in the mid 1990s and funding in FY 2008 ($1.8 billion) was about 25 percent lower in constant 2007 dollars than it was at the peak. The committee was asked to consider priorities under two budget scenarios: one that would require new resources and one that could be achieved through reprogramming of existing funds. Significant new resources would be required for a climate observing system, regional modeling, and user-driven research to support a national climate service. Some new resources could result from entraining additional agencies or agency programs into the restructured climate change research program, or by participating agencies increasing their allocation. The investments of state and local governments in adaptation and mitigation research may also be able to be leveraged to increase the overall research investment. However, these efforts would likely be insufficient to fully implement the priority initiatives.

Under the reprogramming scenario, important adjustments to the program can still be made. The cost to produce the 21 syntheses and assessment reports was about the same as the cost of the last national assessment. Therefore, a national assessment should be within the scope of existing agency funding. Program Office funds could be used to reorganize research around societal issues and to plan critical activities that are not yet funded. Key planning steps include prioritizing climate observations and scoping a national climate observing system and a national climate service. Trades within the program can also be made to expand current activities and advance research on modeling, user-driven research, and adaptation, mitigation, and vulnerability research. For example, a comprehensive research effort on adaptation, mitigation, and vulnerability would require a substantial increase in funding, but since current funding levels directed toward this research are low, the total amount in the initial implementation phase would be relatively small.

Although such reprogramming would be better than business as usual, it would be woefully inadequate for addressing the urgent need to improve our understanding of climate change and satisfy the growing demand for information and analysis to inform action. An inability to meet public expectations would compromise the effectiveness of the new climate change research program. Since the future costs of climate change are expected to greatly exceed the current cost of the federal program, investing now in climate change research should lead to reduced costs for responding, coping with, and adapting to the consequences of climate change. Not investing is a choice we cannot afford to make.

1

Introduction

A CHANGING CONTEXT FOR CLIMATE RESEARCH

Climate change is one of the most important global environmental problems facing the world today. A strong scientific consensus has developed that the observed large warming trend of the late twentieth century will continue unabated in the coming decades and that human activities are the major drivers for many of the observed changes. The United States has been experiencing unusually hot days and nights, heavy downpours, severe droughts, and frequent fires in regions such as California (Karl et al., 2008). More intense hurricanes with the future warming of the tropical north Atlantic are also a potential threat for the United States (Elsner et al., 2008).

Despite international agreements such as the Kyoto Protocol, global consumption of fossil fuels continues to grow about 1.8 percent annually (IEA, 2007), driven by demand for energy both in developed countries, which are responsible for most of the historical accumulation of carbon in the atmosphere, and in emerging economies such as China and India. Globally, CO_2 emissions grew at a record rate of 3.5 percent per year from 2000 to 2007, compared with a rate of 0.9 percent per year from 1990 to 1999 (Global Carbon Project, 2008). World marketed energy consumption is projected to grow by 50 percent from 2005 to 2030 (EIA, 2008b). CO_2 concentrations from fossil fuel burning and other sources are projected to increase from 2005 levels of 379 ppm to

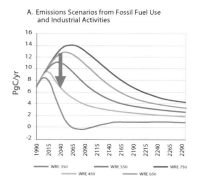

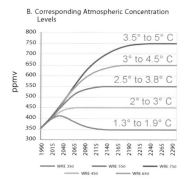

FIGURE 1.1   Illustrative CO_2 emission profiles (A) and corresponding concentrations (B) derived from Wigley et al. (1996) and given in CCTP (2006). The equilibrium surface temperature change associated with steady-state concentrations is shown in red in (B). The surface warming estimates adopt the IPCC (2007a)-recommended climate sensitivity of 3°C warming due to a doubling of CO_2. In addition, they assume that aerosols from air pollution are eliminated and that other greenhouse gases are fixed at 2005 values. SOURCE: Modified from CCTP (2006).

about 440 ppm by 2030 (Figure 1.1), committing the planet to additional warming. These projections are based on estimates that CO_2 emissions in China increased at an annual rate of about 3 to 4 percent during the past 10 years (IPCC, 2007a; IEA, 2007), but a subsequent province-based inventory concluded that emissions actually increased at a higher rate of about 10 to 11 percent (Auffhammer and Carson, 2008). For comparison, total fossil fuel emissions from the United States increased by about 11 percent over the entire 10-year period.[1] Emissions from a number of other developed countries were also higher than agreed-to targets. These disparities between projected and actual emissions underscore the large uncertainties inherent in projecting CO_2 and other greenhouse gas emissions, particularly beyond a decade.

The Intergovernmental Panel on Climate Change (IPCC) projections may have been too conservative in other cases as well. For example, observed increases in surface temperatures and sea level from 1990 to 2007 were in the upper range of IPCC model predic-

---

[1] *http://cdiac.ornl.gov/trends/emis/tre_usa.html.*

tions (Rahmstorf et al., 2007). The retreat of summer Arctic sea ice and snow extent (Déry and Brown, 2007) and melting of the Greenland and Himalayan-Tibetan glaciers (Liu et al., 2006; Kulkarni et al., 2007) may also be larger and faster than predicted. Again, these errors illustrate the large uncertainties in projections of future climate by models used in IPCC and other assessments.

Although the scientific consensus is that the global climate is changing, the research is less conclusive on whether the frequency of abnormal climate events (e.g., prolonged droughts, extensive flooding) will change, how climate change will be manifested regionally, or what impact the changes will have on society. The effects of climate change as well as the vulnerability and resilience of communities and their ability to respond are expected to vary by region (Adger et al., 2007). These effects will not be felt in isolation—the climate is changing against a backdrop of a growing world population and a global economy. At risk is the capacity of the world to provide affordable energy, water, and food to 6.7 billion people. Continuation of the trends of the latter half of the twentieth century, predicted by the IPCC, will introduce natural and social system stresses that will affect public health, economic prosperity, and national security (Box 1.1). Increased greenhouse gas levels have already warmed the planet by 0.8°C and even without further increases, the planet will warm another 0.5°C to 2.5°C, depending in part on future regulation of aerosol emissions (IPCC, 2007a; Ramanathan and Feng, 2008). Planned adaptation, in addition to mitigation, is already becoming necessary.

The public and private sectors are beginning to take actions to adapt to climate change and to mitigate future effects, from shifts toward renewable sources of energy by power companies to greenhouse reduction statutes and policies in California and other states to regional and international carbon trading and offset programs (e.g., Chicago Climate Exchange, European Union's Emission Trading Scheme; Rabe, 2004). Nearly 80 percent of U.S. states have adopted or are preparing climate action plans,[2] and some are taking action to mitigate greenhouse gas emissions, often in partnership with regional efforts such as the Regional Greenhouse Gas Initiative (2005, northeastern states), Western Climate Initiative

---

[2] http://www.perclimate.org/what_s_being_done/in_the_states/action_plan_map.cfm.

(2007), Energy Security and Climate Stewardship Platform for the Midwest (2007), Clean and Diversified Energy Initiative (2004, Western Governor's Association), and the Midwestern Regional Greenhouse Gas Reduction Accord (2007). Foundations are funding hundreds of grants for applied climate change research, much of it dealing with evaluating and informing policy.[3] More than 235 climate-related bills, resolutions, or amendments were introduced in the 110th Congress, twice as many as were introduced in the preceding session,[4] and the Select Committee on Energy Independence and Global Warming was created in the House of Representatives. Authorization for research was a common theme in a number of the bills, including research needed to support decisions on mitigation and adaption (see Appendix A for examples of U.S. legislation under consideration).

It is in this context of larger than predicted climate changes, alarming increases in CO_2 emissions, and decision makers at all levels increasingly willing to respond to such unprecedented developments that we must consider how climate change research should evolve in the United States. A federal science program is needed to comprehend the nature and extent of the climate change threat, to quantify the magnitude of impacts, and to provide a data and knowledge foundation for identifying effective adaptation and mitigation options, with sufficient flexibility to respond to unforeseen problems. Despite these pressing requirements, however, the federal climate change research budget has shrunk from a peak of about $2.4 billion in the mid 1990s to $1.8 billion (in constant 2007 dollars) today.[5]

---

[3] A search of the Foundation Center's directory (*http://fconline.fdncenter.org*) revealed over 300 grants made by almost 50 different private foundations for climate change-related research from 2003 to 2008, totaling nearly $62 million. An assessment by California Environmental Associates identified roughly $200 million of total annual philanthropic funding for climate issues (see *http://www.climate actionproject.com/docs/Design_ to_Win_8_01_07.pdf*).

[4] *http://www.pewclimate.org/what_s_being_done/in_the_congress/ 110thcongress.cfm.*

[5] See *http://www.climatescience.gov/infosheets/ccsp-8/.* Although it is clear that the CCSP budget has declined, the amount is unknown because which activities are included in the program are designated by the participating agencies and vary from year to year (NRC, 2007c). For example,

---

BOX 1.1 Climate Change and U.S. National Security

Climate change is increasingly being discussed in the United States as a national security issue. A number of independent think tanks have identified climate change as a threat to national security (e.g., Busby, 2007; CNA Corporation, 2007). In May 2007, the Senate Committee on Foreign Relations held a hearing on climate change threats from the perspective of the U.S. military.[a] In June 2008, a national intelligence assessment entitled National Security Implications of Global Climate Change to 2030 was presented to the House Permanent Select Committee on Intelligence and the House Select Committee on Energy Independence and Global Warming.[b] The chair of the National Intelligence Council testified that the most significant climate impacts on U.S. national security will be through climate-driven effects on other countries. For example, increasing poverty, food and water shortages, intrastate disputes over water resources, and economic migration could exacerbate political instability in regions such as sub-Saharan Africa, the Middle East, and Southeast Asia. The intelligence assessment, which relied on CCSP results and other published sources, calls for better information on the physical, agricultural, economic, social, and political impacts of climate change at state and regional levels; a better understanding of human behavior; and research to integrate social, economic, military, and political models.

In January 2009, the White House issued a national security presidential directive updating its policy on the Arctic region to account for the effects of climate change, human activity, and altered national policies on homeland security and defense.[c] In the directive, international scientific cooperation—including collaborative research, data collection, and modeling to predict regional environmental and climate change—is seen as vital to promoting U.S. interests in the region. CCSP-sponsored research results and products are likely to be important for implementing the directive.

---

[a] http://foreign.senate.gov/hearings/2007/hrg070509a.html.
[b] Testimony of Thomas Fingar, Deputy Director of National Intelligence for Analysis and Chairman of the National Intelligence Council, before the House Permanent Select Committee on Intelligence and the House Select Committee on Energy Independence and Global Warming, on the National Intelligence Assessment, National Security Implications of Global Climate Change to 2030, June 25, 2008, http://media.npr.org/documents/2008/jun/warming_intelligence.pdf.
[c] White House Memorandum on Arctic Region Policy, National Security Presidential Directive NSPD 66, January 9, 2009.

---

funding to NOAA's laboratories was counted as CCSP beginning in FY 2006, and NASA revised which missions it counted as supporting CCSP goals in FY 2008 (CCSP, 2008).

COMMITTEE CHARGE AND APPROACH

The Global Change Research Act of 1990 established the U.S. Global Change Research Program (USGCRP) to coordinate federally-sponsored research "to understand, assess, predict, and respond to human-induced and natural processes of global change."[6] A new administration in 2001 ushered in the Climate Change Science Program (CCSP), which placed new emphasis on investigating uncertainties and expanded the USGCRP mandate to include research that could yield results within a few years, either by improving decision-making capabilities or by contributing to improved public understanding. The vision for the CCSP is "a nation and the global community empowered with the science based knowledge to manage the risks and opportunities of change in the climate and related environmental systems" (CCSP, 2003). The change of administration in 2009 will likely result in another change in the name and emphasis of the program. In this report, the post-CCSP is referred to as a "restructured climate change research program."

This report is the second of two on the evolution of the CCSP. The first report, *Evaluating Progress of the U.S. Climate Change Science Program: Methods and Preliminary Results* (NRC, 2007c), assessed CCSP progress over the past 4 years (see the Preface for a summary of the findings). This second report identifies future priorities for addressing pressing national and global problems related to climate changes. The charge to the committee was:

> *Task 2. The committee will examine the program elements described in the Climate Change Science Program strategic plan and identify priorities to guide the future evolution of the program in the context of established scientific and societal objectives. These priorities may include adjustments to the balance of science and applications, shifts in emphasis given to the various scientific themes, and identification of program elements not supported in the past. The recommendations will specify which priorities could likely be addressed through an evolution of existing activities or reprogramming, and which would likely require new resources or partnerships.*

---

[6] P.L. 101-606, 104 Stat. 3096-3104 (1990).

The CCSP is organized along scientific themes (e.g., atmospheric composition) or crosscutting issues (e.g., observations) that largely followed the structure of the USGCRP (Appendix B). Such an approach was effective when the main research focus was on understanding how the climate system works. Addressing the research challenges noted above, however, requires a more comprehensive approach that better incorporates and integrates research on natural science, human dimensions, and practical applications (e.g., decision support; see definitions in Box 1.2) to address multiple interactions, feedbacks, and options for action.

To illustrate what is meant by an integrated approach, the committee chose seven examples of climate change issues of importance to society that will have to be addressed in a restructured climate change research program. Examples of such societal issues are illustrated in Figure 1.2. The committee then matched the societal issues with research priorities identified from meetings, workshops, white papers, and peer-reviewed literature. The research and infrastructure (e.g., modeling) needed to address the integrated scientific-societal issues formed the basis for the committee's final list of priorities for a restructured climate change research program. The envisioned research program laid out in this report is ambitious and daunting, but so are the challenges posed by global warming and the potential strategic impacts on our nation.

The climate-energy nexus is at the core of everything discussed in this report. In choosing its priorities, the committee assumed that renewable energy, energy efficiency, and geoengineering and other technologies for mitigating climate change would continue to remain the responsibility of the Climate Change Technology Program (CCTP). Although the committee recognizes that developing mitigation options requires CCSP science—for example, assessments of the environmental impacts of proposed low- and no-carbon energy technologies will undoubtedly be needed—a review of CCTP science needs was beyond both the charge and resources available to the committee.

BOX 1.2 Definition of Terms Used in This Report

Adaptation: Adjustment in natural or human systems in response to climatic stimuli or their effects, which moderates harm or exploits beneficial opportunities

Applications: Activities that use research results to further practical objectives, such as informing the public about regional climate change impacts and supporting decision making

Climate change issues of importance to society: Widely discussed topics that could affect the public's well-being, such as long-term drought

Climate quality observations: Physical or biological observations capable of producing a time series of measurements of sufficient length, consistency, and continuity to determine climate variability and change

Climate services: A mechanism to identify, produce, and deliver authoritative and timely information about climate variations and trends and their impacts on built, social-human, and natural systems on regional, national, and global scales to support decision making

Mitigation: A human intervention to reduce the anthropogenic forcing of the climate system, such as reducing greenhouse gas emissions and enhancing greenhouse gas sinks

Operations: Routine provision of science-based products and services developed to meet specialized needs of stakeholders, either for decision making (e.g., local or regional forecasts) or in support of long-term research (e.g., continuous and systematic measurements of climate variables)

Science: Research aimed at discovering fundamental truths about nature, motivated by either intellectual curiosity or social aims

- **Natural science**: Research on the behavior of the natural (physical-biogeochemical) climate system

- **Human dimensions**: Research drawing on the social, economic, and behavioral sciences and covering human systems drivers of climate change, human systems impacts of climate change, and human systems responses to concerns about or observed effects of climate change

- **Integrated research**: A multidisciplinary/interdisciplinary approach to a particular climate change issue that addresses physical, biological, and human dimensions research and their relationships, interactions, and feedbacks, as well as the research needed to support applications

Stakeholders: Individuals or organizations that generate or use climate information and products, including research scientists; private companies, and nongovernmental organizations in the insurance, agriculture, energy, forestry, transportation, water resources, public health, and emergency response sectors; federal, state, and local government agencies; and policy makers

---

SOURCES: NRC (2004a, 2005b); IPCC (2007c, d).

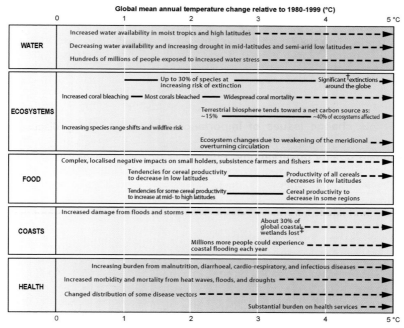

FIGURE 1.2 Examples of societally important issues, in the form of major impacts of climate changes associated with increasing global temperatures. The left side of the text indicates when impacts (black lines) begin and the dashed arrows show their continuation with rising temperature. NOTE: † Significant is defined here as more than 40 percent. ‡ Based on an average rate of sea level rise of 4.2 mm/year from 2000 to 2080. SOURCE: Adapted from IPCC (2007c), Figure SPM2, Cambridge University Press. Used with permission.

ORGANIZATION OF THE REPORT

This report lays out an approach for integrating scientific and societal objectives and identifies priorities for a restructured climate research program. Chapter 2 presents examples of seven scientific issues of importance to society and the integrated research needed to address them. The committee's process for identifying the research needs is described in Appendix C. The starting point was the gaps and weaknesses identified in the NRC (2007c) report *Evaluating Progress of the U.S. Climate Change*

Science Program: Methods and Preliminary Results (Preface) and discussion papers on research priorities in the human dimensions (Appendix D) and natural science (Appendix E) prepared by the Committee on the Human Dimensions of Global Change and the Climate Research Committee, respectively. These priorities were vetted at two stakeholder workshops by individuals listed in Appendix F, and the final ones were chosen by the committee. Chapter 3 discusses the current gaps, shifts in emphasis, and future priorities for a restructured climate research program, along with the organizational and resource implications for implementing them. Finally, biographical sketches of committee members and a list of acronyms and abbreviations appear in Appendixes G and H, respectively.

2

Restructuring the Climate Change Science Program

S ocieties' ability to respond to climate change depends in part on the magnitude and speed of changes in the climate system and on the resilience of human and environmental systems in the face of these changes. Air and ocean temperatures are increasing, resulting in widespread melting of snow and ice and rising sea levels. This global warming has been occurring over the past century, but has greatly accelerated in the past few decades, driven by the addition of greenhouse gases, especially CO_2, to the atmosphere at an ever increasing rate. A warming in excess of 3°C is possible (cf., Figure 2.1) and could push components of the climate system past various tipping points (e.g., Schneider and Mastrandrea, 2005; Lenton et al., 2008), including the possible loss of the major ice sheets and glaciers. The bell-shaped curve of the warming with a wide range of 1.5°C to 4.5°C and a "fat tail" shown in Figure 2.1 illustrates the large uncertainty in our understanding of the response of the climate system to human perturbation. It also suggests that we cannot entirely dismiss the possibility of irreversible changes in the way Earth's climate operates and how human and ecological systems respond.

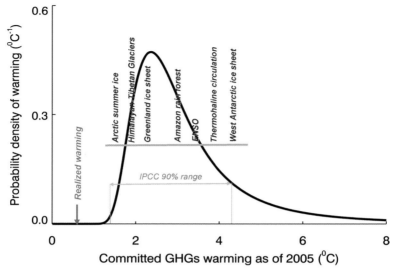

FIGURE 2.1 Probability distribution of the predicted increase in global mean surface temperature due to a 3 Wm^{-2} radiative forcing from increases in greenhouse gases from preindustrial times to 2005. The probability density of the expected warming adopts the IPCC (2007a) climate sensitivity of 3°C warming due to a doubling of CO_2, with a 90 percent confidence level of 2°C to 4.5°C warming. The realized warming is the warming from 1750 to 2005 that has been attributed to greenhouse forcing. Because of the small amount of warming that has been realized to date and the presence of strong cooling by aerosols, temperature increases above 2°C are likely not imminent but could be very large before the end of the century. The temperature thresholds for various climate tipping points are marked by the blue words. The ranges, taken from Lenton et al. (2008), are not shown, but are 0.5°C to 2°C for the melting of Arctic summer sea ice; 1°C to 2°C for radical shrinkage of the Greenland Ice Sheet and 3°C to 5°C for shrinkage of the West Antarctic Ice Sheet; 3°C to 4°C for the dieback of the Amazon rain forest due to drastic reductions in precipitation; 3°C to 6°C for persistent El Niño conditions; and 3°C to 5°C for a shutoff in the North Atlantic deep water formation and the associated thermohaline circulation. The tipping point of Himalayan-Tibetan glaciers is based on the IPCC (2007a) finding that these glaciers may suffer drastic melting when warming exceeds 1°C to 2°C above preindustrial levels. SOURCE: Ramanathan and Feng (2008).

What measures society should, can, and will apply to slow the growth of greenhouse gases and/or reduce the dangers posed by the expected large climate system changes are still far from settled. Changes in greenhouse gas emissions reflect behavioral patterns, energy consumption, population growth, and societal responses to climate change. These changes are happening in the context of complex socioecological systems in which nature and society are mutually dependent and are constantly affecting one another (positively and negatively) across space and time (Folke, 2006). The fundamental dilemma faced by policy makers is how to forge effective strategies both to mitigate further climate change and to adapt to the changes already under way, in view of the uncertainties in our knowledge about how climate affects humans and vice versa and of the political difficulties of taking costly action now for benefits that accrue in the future. The fat tails of the distribution of climate sensitivity (Figure 2.1), rather than the average, may drive the economic trade-offs associated climate change (Weitzman, 2009). Policy and decision makers must have better information that meets their needs (NRC, 2009).

Improving understanding of the interactions and feedbacks of the physical climate system with human and environmental systems, improving predictions of longer-term causes and trends, and preparing the nation for future climate changes are grand challenges. They are particularly difficult to tackle if we do not understand the system as a whole. Under the Climate Change Science Program (CCSP), much has been learned about components of the natural climate system, including the composition of the atmosphere, the water and carbon cycles, and changes in the land surface (NRC, 2007c). It is now time to take a more holistic approach and integrate across natural and social science disciplines and across the science and policy worlds to find solutions to climate change-related problems that are of major concern to society.

This chapter provides seven examples of societal issues that motivate the need for an integrated approach to the research program. Two are current issues stemming from changes in the climate system (weather and climate extremes, sea level rise and melting ice) and five focus on impacts of climate change (availability of freshwater, agriculture and food security, managing ecosystems, human health, and impacts on the economy of the

United States). The examples connect societal issues widely recognized as essential to the well-being of the planet with high-priority science and application needs. Although not a comprehensive list, they show how the CCSP could be organized to yield both improved understanding of the climate system and the knowledge foundation needed to support sound decision making.

EXTREME WEATHER AND CLIMATE EVENTS AND DISASTERS

Extreme (severe) weather and climate events are the most visible manifestations of climate-related hazard. In the worst cases, such extreme events interact with socioeconomic, political, and ecological factors (e.g., food and water supply) to create economic or health disasters (Wisner et al., 2004). Especially at risk are the poor, uneducated, very old or very young, and the sick. How society deals with extreme weather events today provides an analog for understanding our vulnerability to hazard in a changing climate (Adger et al., 2003). The impact of climate-related hazard depends on two factors: (1) the level of exposure to the danger (e.g., storms, heat waves, droughts) and (2) the capacity of the vulnerable party to respond, cope, and adapt (Wisner et al., 2004; Tompkins et al., 2008). For example, Hurricane Mitch killed thousands when it struck Honduras in 1998, but had a much less devastating impact on Florida (Glantz and Jamieson, 2000). The reasons for the disparate consequences relate both to the changing nature of exposure (Mitch started as a category 5 hurricane in the Caribbean and ended as a tropical storm in Florida) and to the high levels of poverty in Honduras, where many died because they did not have the means to flee or to "ride out the storm."

Even in a country as wealthy as the United States, the growing frequency and cost of climate-related disasters have taken a toll. In the 1990s there were 460 presidential disaster declarations, nearly double the number of the previous decade, and 498 declarations were made from 2000 to October 2008.[1] Of the 62 weather-related disasters that cost more than $1 billion between 1980 and 2004, one-quarter hap-

---

[1] *http://www.fema.gov/news/disaster_totals_annual.fema.*

pened after 2000 (DOC, 2005, cited by Burby, 2006: 172). Hurricane losses since 1990 have risen dramatically, both in absolute terms and as a fraction of gross domestic product (Nordhaus, 2006), mostly because of increases in the population and the value of assets in exposed coastal regions (Pielke et al., 2008). Higher costs can be expected as climate continues to change (IPCC, 2007a).

Research on climate vulnerability has identified many factors, both positive and negative, that shape the level of exposure and sensitivity of people and settlements (Eakin and Luers, 2006; see also Backlund et al., 2008; Gamble, 2008; Savonis et al., 2008). For example, changing demographics in U.S. coastal areas have likely increased overall vulnerability to storm-related flooding and damaging winds. Not only are more people living permanently (rather than seasonally) on coasts, they also are older (retirees), more racially and ethnically diverse, and more likely to have low-wage jobs (Cutter and Emrich, 2006). Approximately half of the U.S. population, 160 million people, lives in a coastal county (Gamble, 2008). By 2050, 86 million people in the United States will be 65 or older and potentially more sensitive to the effects of heat waves and flooding. Managing this vulnerability requires both short-term actions to prevent disasters and assist recovery efforts (e.g., evacuation; supply of clean water, shelter, and food; reconstruction of infrastructure) and longer term structural reforms to reduce people's vulnerability to disasters (e.g., land-use regulation; Lemos et al., 2007).

By definition, extreme events occur infrequently, typically as rare as, or rarer than, the top or bottom 10 percent of all occurrences. A relatively small shift in the mean climate, caused by human activities or natural variability (e.g., changes in atmospheric circulation associated with the El Niño/Southern Oscillation [ENSO] phenomenon), can produce a larger change in the number of extremes. In a changing climate system, some extreme events will be more intense, some will occur more frequently, and others will occur less frequently (Karl et al., 2008). Yet building codes and insurance premiums are based in part on the occurrence of extreme events in the past.

Over the past few decades, the number of heat waves and warm nights has increased in the inhabited continents, while cold days, cold nights, and days with frost have become rarer (Figure 2.2). The

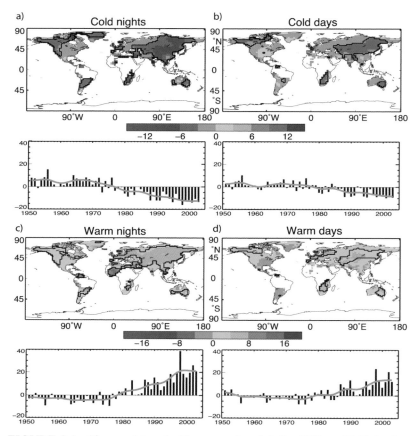

FIGURE 2.2 Observed trends (days per decade) for 1951 to 2003 in the frequency of extreme temperatures, defined on the basis of 1961 to 1990 values, as maps for the 10th percentile, (a) cold nights and (b) cold days; and 90th percentile, (c) warm nights and (d) warm days. Trends were calculated only for grid boxes that had at least 40 years of data during this period. Black lines enclose regions where trends are significant at the 5 percent level. Below each map are the global annual time series of anomalies (with respect to 1961 to 1990). The orange line shows decadal variations. Trends are significant at the 5 percent level for all the global indices shown. SOURCE: From Trenberth et al. (2007), FAQ 3.3, Figure 1, Cambridge University Press. Adapted from Alexander et al. (2006).

United States has experienced fewer severe cold episodes over the past decade than for any other 10-year period in the U.S. historical climate record, which dates back to 1895 (Kunkel et al., 2008). One of the adverse consequences of warmer winters (along with prolonged drought stress and forest management practices) is the spread of the pine bark beetle, which has decimated forests in the western United States (Negrón et al., 2008).

Global warming also influences changes in precipitation. Air holds more water as it warms (Dai, 2006; Santer et al., 2007), resulting in more moisture for storms and thus heavier rainfalls or snowfalls and greater potential for flooding. For the contiguous United States, statistically significant increases in heavy (upper 5 percent) and very heavy (upper 1 percent) precipitation have been observed over the past three decades (Kunkel et al., 2008), and heavy rain events are contributing more to the total precipitation (Klein Tank and Können, 2003; Groisman et al., 2004; Alexander et al., 2006).

At the same time, warmer air leads to greater evaporation and surface drying in some areas and thus contributes to drought and increased risk of wildfires. Over the past several decades, drought has increased, especially in Africa, southern Asia, the southwestern United States, Australia, and the Mediterranean region (Figure 2.3). The extent of very dry land across the globe has more than doubled since the 1970s (Dai et al., 2004) as a result of decreases in precipitation and the large surface warming. Like other climate-related impacts, the impacts of drought depend on a combination of stressors at different scales (Wilbanks et al., 2007). For example, populations already stressed by poverty, warfare, or AIDS are more vulnerable to drought (see "Freshwater Availability," below). Understanding how these stressors combine and interact is essential for informing policy.

Intense extratropical cyclones can produce extremely severe local weather, such as thunderstorms, hail, and tornadoes. Such storms appear to be increasing in number or strength (e.g., Wang et al., 2006), and their tracks have been shifting northward in both the North Atlantic and North Pacific over the past 50 years (e.g., Gulev et al., 2001; McCabe et al., 2001). Climate models project these storms to be more frequent over the next century, with stronger winds and higher waves (Meehl et al., 2007).

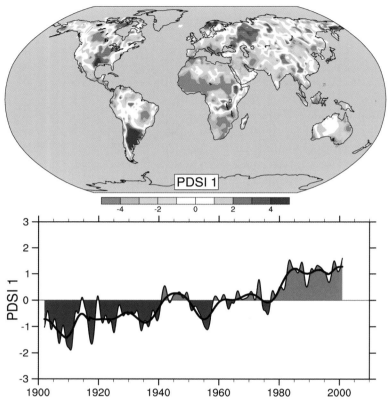

FIGURE 2.3 (Top) Spatial pattern of drought for 1900 to 2002, as represented by the monthly Palmer Drought Severity Index (PDSI), which measures the cumulative deficit (relative to local mean conditions) in surface land moisture. The lower panel shows how the sign and strength of this pattern has changed since 1900. Red and orange areas in the top panel are drier (wetter) than average and blue and green areas are wetter (drier) than average when the values shown in the lower plot are positive (negative). The smooth black curve shows decadal variations. Widespread drought is increasing in Africa, especially in the Sahel, while some regions are getting wetter, especially in eastern North and South America and northern Eurasia. SOURCE: Trenberth et al. (2007), FAQ 3.2, Figure 1, Cambridge University Press. Adapted from Dai et al. (2004).

Of all extreme events, however, tropical cyclones cause the greatest property damage (e.g., Box 2.1), and so any changes in their frequency and intensity are vital to residents who live in their paths, state and local disaster preparedness organizations, and the insurance industry (Murnane, 2004). The number of tropical storms and hurricanes affecting the United States fluctuates from decade to decade, and data uncertainty is larger prior to 1965, when the satellite era began (Gutowski et al., 2008). Nonetheless, it is likely that the annual number of tropical storms and hurricanes in the North Atlantic has increased over the past 100 years, although there appears to be no trend in the proportions of major hurricanes or in overall intensity (Holland and Webster, 2007). When multiple storms hit the same region, as happened in Florida and Louisiana in 2005, communities have little time for recovery and resilience building.

Since about 1970, and likely since the 1950s, Atlantic tropical storm and hurricane destructive potential has increased (Emanuel, 2005, 2007). The destructive potential is strongly correlated with tropical Atlantic sea surface temperatures. Model simulations suggest that for every 1°C increase in tropical sea surface temperature, core rainfall rates will increase by 6 to 18 percent and the surface wind speeds of the strongest hurricanes will increase by about 1 to 8 percent (Gutowski et al., 2008). Other changes in the climate system (e.g., higher sea level) as well as growing populations and development in coastal zones will worsen the impacts of hurricanes and the associated storm surges and beach and wetland erosion.

Research Needs

Because humans both contribute to extreme weather and climate events and suffer from their consequences, research is needed to understand the underlying physical and human processes and their interactions, feedbacks, and impacts, as well as to meet the information needs of stakeholders developing warning systems and response and adaptation options. For example, states need improved understanding and prediction of storm events with the potential to generate major regional flooding (CDWR, 2007). Research is also needed on how to account for changing socioeconomic conditions, including adaptation over time, to improve our understanding of

BOX 2.1 Hurricane Katrina

Hurricane Katrina was one of the worst disasters in U.S. history and offers important lessons on how U.S. coastal regions may be vulnerable to potential increases in hazard related to future climate change. The category 3 storm, which hit New Orleans in August 2005, caused $81 billion in total damage and $40.6 billion in insured losses. On the northern Gulf coast, 1.2 million people were evacuated from their homes and 1,833 people were killed, directly or indirectly. In its wake, 43 tornadoes touched ground in Florida, Georgia, Alabama, and Mississippi. The different levels of vulnerability of individuals and communities became painfully clear in the aftermath of the hurricane. Preventing similar disasters will require research from a wide range of disciplines, including atmospheric physics, biology, sociology, engineering, political science, economics, anthropology, and psychology (Gerber, 2007). However, science alone will not solve the problem if integrated approaches and better communication, disaster management, and policy capacity are not in place (e.g., Waugh, 2006).

FIGURE People's ability to flee or to recover from the negative impacts of Hurricane Katrina revealed the many social, physical, structural, and political dimensions of extreme weather or climate events. (Left) Vehicles leave New Orleans ahead of Hurricane Katrina on August 28, 2005. SOURCE: AP Photo/Bill Haber. (Right) Thousands wait to be evacuated from the Superdome in New Orleans, September 2, 2005. SOURCE: REUTERS//David J. Phillip/Pool.

SOURCE: Weather Channel,
http://www.weather.com/newscenter/topstories/060829katrinastats.html.

losses associated with climate extremes. Specific research needs include the following (Gamble, 2008):

•   Improved understanding of climate thresholds and vulner-abilities, impacts, and adaptive responses (including adaptation limitations) in a variety of different local contexts around the country
•   Improved understanding of population changes and migra-tion, especially in areas of high vulnerability
•   Improved understanding of vulnerable populations (e.g., the urban poor, native populations on tribal lands) that have limited capacities for responding to climate change. The results are key inputs to adaptation research that addresses social justice and environmental equity concerns

Given high uncertainties regarding climate impacts, it may make sense to focus more on building adaptive capacity than on developing specific adaptation options for different types of extreme events (Pielke, 2007). Whereas adaptation is local, ways to build adaptive capacity can be generalized across individuals, communities, and countries (Eakin and Lemos, 2006). Research is also needed on the types of incentives that will encourage adaptation (Christopolos, 2008).

Decision support tools are needed by disaster management agencies, first responders, city planners, and others responsible for hazard mitigation and management. Examples of the science needed to manage flood risk in the context of climate change include (CDWR, 2008b):

•   Updated flood frequency analyses of major rivers and streams
•   Studies of forecast-based operations for major reservoirs
•   Analysis of the costs and benefits of adjusting state water supply and flood control infrastructure to accommodate climate variability
•   Assessment of innovative techniques for improving flood risk evaluation, including use of paleoflood reconstructions

Finally, much of the research on the natural climate system and human contributions and responses relies on a good observational record that enables trends in climate, including extremes, to be discerned with a high level of confidence. However, large areas of the world, even large parts of North America, are underobserved. Moreover, most observations used for climate purposes are obtained from weather observing networks, although these data often reflect nonclimatic changes from station relocations, land-use changes, instrument changes, and observing practices that have varied over time. Only a few countries have developed true climate observing networks that adhere to the Global Climate Observing System (GCOS) climate monitoring principles.[2] Research using observations for which nonclimatic changes have been removed, therefore, would provide a better understanding of climate system variability in extremes.

Because of the presence of multidecadal modes of variability in the climate system, an understanding of natural and human effects on historical weather and climate extremes is best achieved through study of very long (century-scale) records. For many of the extremes discussed above, including temperature and precipitation extremes, storms, and drought, long-term, high-quality, homogeneous records are not available. Particular requirements to further improve our understanding and detection of changes in weather and climate extremes include the following (Easterling et al., 2008):

• Research on how to quantify uncertainty in homogeneity-adjusted climate datasets, and the best adjustment methods
• Continued development and maintenance of high-quality climate observing systems that adhere to the GCOS climate monitoring principles (e.g., U.S. Climate Reference Network[3]), including open exchange of data so more comprehensive analysis products can be produced
• Collection of higher frequency data, such as hourly precipitation
• Collection of socioeconomic observations to inform impact, vulnerability, and adaptation research (e.g., cost-benefit data to analyze adaptation options; data on social networks, preferences, and adaptation resources and institutions; vulnerability indicators)

---

[2] *http://www.wmo.int/pages/prog/gcos/documents/GCOS_Climate_Monitoring_Principles.pdf.*
[3] *http://www.ncdc.noaa.gov/crn.*

• Analysis of long-term observations by multiple, independent experts to improve confidence in detecting past changes

• Creation of annually-resolved, regional-scale reconstructions of the climate for the past 2,000 years to improve our understanding of regional climate variability

• High-temporal-resolution data from climate model simulations to improve understanding of potential changes in weather and climate extremes

SEA LEVEL RISE AND MELTING ICE

The reconstructed record of global sea level (1870 to 2001) reveals an average increase of 1.7 ± 0.3 mm per year (Church and White, 2006), primarily as a result of expansion of warming seawater and discharge of ice from alpine glaciers, ice caps, and the Greenland and Antarctic ice sheets to the oceans. Although the rate of sea level rise varies on decadal scales, over this observational period global sea level exhibited an acceleration of 0.013 ± 0.006 mm yr^{-2} (95 percent confidence; Church and White, 2006). Since 1993, tide gauge and altimetry data confirm the rate of sea level rise to be ~3 mm per year, although this rate was also attained briefly around 1950 and 1970.

This recent acceleration is driven in part by increased thermal expansion and the melting of nonpolar glaciers (Meier et al., 2007). Increased ice discharge from Greenland also plays an important role. Although measuring ice discharge and ice sheet mass balance is challenging (Cazenave and Nerem, 2004), available evidence suggests about a fourfold increase in Greenland ice discharge from 1993 to 2003 relative to the 1961 to 2003 period (IPCC, 2007b, Chapter 5). Observations using advanced technologies point to accelerated ice losses since 1993 ranging from about 60 percent (1993 to 1998; Krabill et al., 2004) to a threefold increase (1993 to 1998 relative to 1998 to 2004; Thomas et al., 2006).

The impact of melting sea ice on polar bear habitat is becoming iconic,[4] but melting ice is also affecting human settlements. For example, the Inuit people of North America are having to change hunting and fishing practices and travel routes, and their cultural traditions and health are being adversely affected (Hassol, 2004). Rising seas added to high tides and storm surges will have profound effects on the built environment (e.g., Box 2.2) and ecosystems in coastal areas. Intrusion of saltwater will affect groundwater quality and supplies (Backlund et al., 2008), and is one of the most severe threats to long-term agricultural sustainability in the Pacific Islands (Shea et al., 2001). Higher storm surges will disrupt sewer systems and water treatment facilities and promote rapid barrier island migration or segmentation, disintegrating wetlands (CCSP, 2009). Coastal and near-shore ecosystems such as coral reefs, mangroves, and sea grass communities as well as the coastal fisheries they support are particularly vulnerable to rising sea levels and increased storm surges.

Predictions of how fast and how much sea level might rise are hampered by the scarcity of observations. Recent observations of increased ice discharge from Greenland (Rignot and Kanagaratnam, 2006; Howat et al., 2007), West Antarctica (Thomas et al., 2004; Rignot et al., 2008), and the Antarctic Peninsula (Scambos et al., 2004) were not included in Intergovernmental Panel on Climate Change (IPCC) projections. Thus, the IPCC projection of 0.18 to 0.59 meter of sea level rise by 2100 is likely an underestimate (IPCC, 2001b, Technical Summary; Rahmstorf, 2007; Pfeffer et al., 2008). Two recent studies using different approaches concluded that an increase of 1 meter by 2100 lies well with projected ranges (Rahmstorf, 2007; Pfeffer et al., 2008). A global sea level rise of 1 meter would affect 145 million people (most in Asia) at a cost of nearly 1 trillion U.S. dollars (IPCC, 2007c, Table 6.12). It would also inundate 65 percent of the coastal marshlands and swamps in the contiguous United States (Backlund et al., 2008), affecting habitat quality and triggering rapid nonlinear ecological responses (Burkett et al., 2005). Figure 2.4 shows the global and a local (San Francisco) area expected to be inundated by a 1 meter rise in sea level.

---

[4] See news stories such as *http://environment.newscientist.com/channel/ earth/climate-change/dn11656* and assessments of the population status of polar bears, such as Schliebe et al. (2006).

**BOX 2.2 Increasing the Adaptive Capacity of
Transportation Systems on the Gulf Coast**

The Gulf Coast is one of the most climate-vulnerable regions in the United States. It is also one of the most critical for energy security since approximately two-thirds of all U.S. oil imports and 90 percent of domestic oil and gas extracted from the outer continental shelf are transported through this region (Potter et al., 2008). The oil and gas transportation networks as well as the regions' complex web of roads, airports, and waterways are vulnerable to sea level rise (and also to warmer temperatures, increased storm activity, and changed precipitation patterns; see Savonis et al., 2008). A sea level rise of 2 to 4 feet would place 27 percent of the major roads, 9 percent of the rail lines, and 72 percent of the ports at or below 4 feet in elevation at risk, despite protective structures such as levees and dikes (Potter et al., 2008). Because the planning time frame of transportation managers is around 20 to 30 years, important decisions that will shape the region's adaptation options for the future are being made today. Although transportation managers are accustomed to planning under high levels of uncertainty (e.g., future travel demand, vehicle emissions, revenue forecasts, seismic risks) and environmental pressure, better climate change-related knowledge (e.g., levels of exposure, vulnerability, resilience) are necessary to develop robust adaptation options. Research needs of interest to decision-makers include integrated climate data and projections, risk analysis tools, and region-based analyses.

Sea levels will change noticeably only over decades, but such changes will continue for many centuries into the future. Knowing how much regional and local sea level is likely to rise would help improve the design and implementation of cost-effective measures to protect against coastal inundation, salinization of groundwater and estuaries, enhanced erosion, and ecosystem losses and for managing long-life infrastructure such as nuclear power plants. For example, Mount and Twiss (2005) estimated that it would cost at least $1 billion to raise the California Central Valley levees just 0.15 meter. Sea level rise will have a greater impact in areas that are subsiding or that have gently sloping shorelines. The mid-Atlantic coast of the United States is an excellent example of a region with high potential for enhanced damage due to storm surges associated with extreme weather events (hurricanes, nor'easters; Najjar et al., 2000). A recent study (Kleinosky et al., 2007) highlights the vulnerability and increased risk of damage for 10 cities in the Hampton Roads, Virginia area due to hurricane storm surges superimposed on sea level rise, population growth, and poorly planned development.

FIGURE 2.4 Areas potentially vulnerable to inundation as a result of a 1 meter rise in sea level, which is within the range expected by many scientists by the end of this century. The red areas in the top image show the global distribution of these low-lying shorelines. SOURCE: National Aeronautics and Space Administration, *http://www.nasa.gov/topics/earth/ tipping_points_hiresmulti_prt.htm*. The light blue areas in the bottom figure depict low-lying areas within the San Francisco Bay area, California, based on U.S. Geological Survey elevation data and imagery from the National Agriculture Imagery Program. SOURCE: San Francisco Bay Conservation and Development Commission.

Research Needs

The threat of rising sea level raises a number of questions that cannot be answered with our current level of understanding. Among the most important are: At what degree of warming will the ice sheets of Greenland and West Antarctica be drastically affected? How will their marine-terminating glaciers and ice streams respond to warmer conditions? What volume of land-based ice might be discharged into the oceans and how rapidly? The controls on glacier flow are dominated by ice dynamical processes that are nonlinear (Howat et al., 2007), raising the possibility that glaciers and ice streams may become so unstable (pass a tipping point) that they will begin to rapidly discharge ice until a new steady state or equilibrium condition is achieved. Once large ice streams and marine-terminating glaciers begin to move, they cannot be stopped by any form of intervention.

The response of large ice sheets to warmer climate conditions has so far been difficult to quantify, model, and predict (Alley et al., 2005; IPCC, 2007b, Technical Summary; Rahmstorf, 2007). The controls on ice flow (dynamics), including the possible influence of meltwater and basal lubrication on glacier discharge, are poorly understood, in part because of scanty observations (Das et al., 2008; Joughin et al., 2008). After the breakup of the Larsen B Ice Shelf on the eastern side of the Antarctic Peninsula, the affected outlet glaciers began to flow two to six times faster, whereas those flowing into the remaining intact parts of the ice shelf did not accelerate (Scambos et al., 2004). Projections of how these ice sheets are likely to respond requires the development of coupled ice sheet–outlet glacier–ocean models that can be nested within global climate system models. In situ data are needed on mass balance components (precipitation, sublimation, blowing and drifting snow), changes in glacier dynamics and subglacial drainage systems, and the thermodynamic interactions of marine-terminating glaciers (Holland et al., 2008) and ice shelves buttressing land-based ice with warm water intruding underneath. Remotely sensed observations (e.g., laser altimeter, synthetic aperture radar [SAR], gravity field differences) are needed to understand the drivers of mass balance changes. Repeat SAR images make it possible to estimate the volume of ice discharge per unit time. Realistic projections of sea level rise demand better ice sheet models and, although progress has been made, no model including all relevant forces yet exists (Alley et al., 2005), nor does one appear imminent.

Despite the predicted negative impacts, U.S. coastal policy often does not take sea level rise into consideration (CCSP, 2009). Research priorities should focus on tools, datasets, and land management information to support and promote sound coastal planning, including better data and resources provided via platforms (e.g., geographic information systems) that improve their usability by decision makers. The research should also link physical vulnerability with economic analysis, planning, and assessment of adaptation options. Specific research needs include the following:

• Understanding of increased risks of and damages from coastal storm surge flooding
• Developing risk management approaches for coastal development and local land-use planning
• Developing "planned retreat" strategies, such as the demolition of large structures near the shore if sea level rises by a specified amount (Titus, 1990) or the prohibition of reconstruction of coastal property severely damaged by repeated flooding (Yohe and Neumann, 1997), or coastal protection strategies that factor in sea level rise and climate change, such as those planned in the Netherlands.[5]

FRESHWATER AVAILABILITY

Climate change poses a grave threat to the availability of freshwater in the United States and around the world. Large populations concentrated in cities and suburbs as well as our entire agricultural base are dependent on, and accustomed to, safe, reliable sources of freshwater. The availability of freshwater involves both supply and demand. Already there is growing demand (e.g., by people, agriculture, industry), unequal distribution (UNDP, 2006), and declining sources (Figure 2.5; IPCC, 2007a; Bates et al., 2008). On the supply side, the availability of freshwater depends not only on the global water cycle, which describes the flows and storage of water in the natural

---

[5] *http://www.deltacommissie.com.*

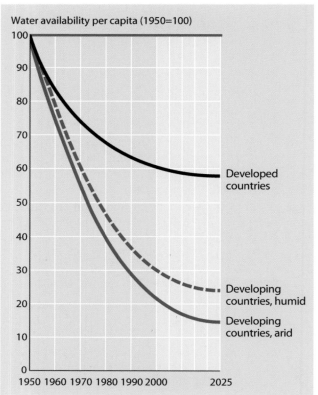

FIGURE 2.5 Declining water availability per capita from 1950, projected to 2025. Developing countries in arid regions are expected to be hardest hit by water scarcity, but even developed countries can expect significant reductions in water availability. SOURCE: UNDP (2006). Adapted from Pitman (2002), data copyright World Bank.

system (precipitation, lakes, river flow, groundwater, snowpack, glaciers, and water vapor), but also on our technical capacity to store and adapt freshwater systems to societal needs (e.g., by building dams and canals or by developing restoration and clean up technologies). On the demand side, freshwater availability depends on governance (individuals, institutions, communities, and organizations), behavior, and the values shaping water use and sustainability (e.g., consumption, conservation, valuation and equitable distribution). Both supply and demand are expected to be affected by a changing climate.

How much change has already occurred and how much is likely to occur in the future is uncertain because of the large natural variability of basic components of the water cycle (e.g., precipitation, stream flow) and the difficulty of predicting many of the social and behavioral processes affecting water (e.g., consumption, conservation). For example, climate is potentially a transformation factor in water governance (Bates et al., 2008). As with many climate variables, uncertainty in both past and projected trends is smallest for large (e.g., global) spatial averages, and largest at the regional scales (e.g., river basins and groundwater reservoirs), where the water cycle most directly affects society and where water managers most need information (Beller-Simms et al., 2008; Lemos, 2008).

Average global precipitation and evaporation are expected to increase, based on theory and confirmed by many models. The model projections of changes in precipitation are not uniform around the globe; generally the increase in precipitation is expected to be most pronounced in the extra tropics, accompanied by a drying of the tropical land from 10S to 30N (Bates et al., 2008). Some observational analyses have suggested that precipitation increase associated with increasing temperatures may be underestimated by current models (Lambert et al., 2008).

The physical processes involved in precipitation and evaporation are complex, involving dynamic processes that occur over length scales smaller than those resolved by climate models, ranging from aerosols to clouds to weather systems. Radiation budgets are affected by both scattering aerosols such as sulfates and absorbing aerosols such as black carbon, which intercept sunlight before it reaches the surface (IPCC, 2007b, Chapter 2). The reduction of sunlight at the ground leads to a decrease in evaporation and a corresponding decrease in precipitation. Aerosols can also nucleate cloud drops and influence rainfall patterns locally and regionally (Rosenfeld et al., 2008). Changes in spatial gradients of sea surface temperatures, due to natural or anthropogenic forcing, also have a major influence on continental precipitation (Box 2.3). The socioeconomic and political processes are also complex and feedbacks among water access, consumption, markets, ecosystems services, equity and gender distribution, security, development, and health are not well understood (UNDP, 2006).

BOX 2.3 Sahel: Drought of Unprecedented Severity

The Sahel region of Africa borders the Sahara Desert and is an area of low rainfall, frequent drought, and limited natural resources (top two figures). The Sahel region as well as the rest of western Africa face major challenges arising from climate variability and the effects of predicted climate changes on food production, freshwater availability, and desertification. The last Sahelian drought from the early 1970s to the mid 1980s is among the worst on record and left about 100,000 dead and close to a million on food aid (Wijkman and Timberlake, 1984).

FIGURE Desert landscape in Mali. SOURCE: Romano Cagnoni/Peter Arnold Inc.

Scientists are still debating the causes of the devastating drought, but the current consensus is that the primary forcing term is decadal-scale changes in ocean temperatures. In particular, warmer temperatures in the Indo-Pacific warm pool and a combination of cooler-than-normal North Atlantic temperatures and warmer-than-normal South Atlantic temperatures are emerging as the dominant factors. Greenhouse warming can account for the warmer Indo-Pacific warm pool temperatures and aerosol cooling can account for the cooler-than-normal North Atlantic sea surface temperatures. But these human forcing terms by themselves cannot account for the amelioration of the drought. Natural variations in Atlantic and Indo-Pacific sea surface temperatures (e.g., ENSO induced) have to be invoked. Scientific uncertainty is hampering our ability to predict future changes in this vulnerable region of sub-Saharan Africa. For example, simulations by two reputable climate models in the United States disagree on even the sign of the changes (bottom figure): one predicting a

40 to 60 percent decrease in rainfall by 2100 and the other predicting a 20 percent increase!

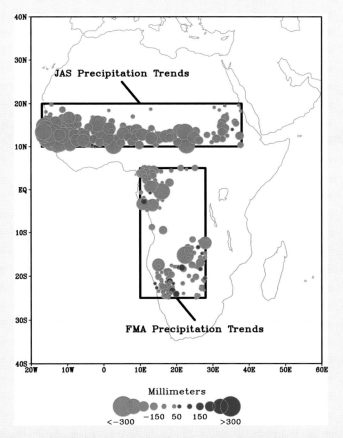

FIGURE Observed precipitation trends from 1950 to 1999. SOURCE: Hoerling et al. (2006). Copyright 2006 American Meteorological Society.

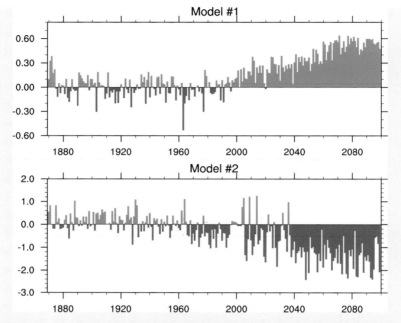

FIGURE Simulated time series of rainfall departures over the Sahel for July through September 1870 to 2099 from two different climate models. Reference climatology is 1870 to 1999. Both models were forced with estimated greenhouse gas and aerosol changes through 1999 and with the SRES A1B emissions scenario (IPCC, 2007b) thereafter. SOURCE: Adam Phillips, NCAR, based on results reported in Hoerling et al. (2006).

The great Sahelian drought illustrates the many factors that influence the impact of water scarcity on ecosystems, communities, and social groups. Twenty years after the drought, research focusing on the interactions between environmental degradation, socioeconomic transformation, and climatic change has painted one of richest pictures of vulnerability and adaptation in the less developed world (Batterbury and Warren, 2001; see also the special issue of *Global Environmental Change*, **11**, 2001). By examining social, political, and environmental change together, this research challenged well-established myths about desertification, competition between human settlements and livestock for land and water, and migration, and showed how large-scale processes at the global level (greenhouse warming and anthropogenic pollution) connect with local-scale processes (livelihood adaptation and local knowledge) to produce drought.

SOURCES: Zeng (2003) and Hoerling et al. (2006).

Scientists are able to predict global average trends—the first-order response of the hydrological cycle to a warming planet. However, the regional changes that are most important for planning water resources are poorly understood and less predictable. For example, changes in the distribution of temperature and water vapor in the atmosphere, as well as in circulation patterns, will change the amount and type of precipitation (snow or rain) that falls across the United States. Coupled with temperature and radiation changes at the land surface, such direct and indirect effects make prediction of critical resources such as the depth of seasonal snowpack in the Rockies difficult. Similarly, for areas dependent on groundwater aquifers, recharge is sensitive not just to total precipitation, but also to changes in storm climatology (intensity, duration, and intermittency of storms, all of which change with climate warming) as well as near-surface weather parameters (e.g., air temperature, humidity; Levine and Salvucci, 1999; NRC, 2004b). Statistical analyses of hydrological and meteorological records have found evidence that such key aspects of the water cycle are changing. For example, observations around the world indicate an increase in frequency of intense rainfall. Models suggest this intensification will increase in the coming decades (Figure 2.6), leading to increased incidence of diseases such as diarrhea (UNDP, 2006), more flooding (Backlund et al., 2008), and, ironically, to possibly less recharge.

The IPCC Fourth Assessment predicts both increases (wet regions get wetter) and decreases (dry regions get drier) in annual average river runoff and water availability, as well as changes in the extent of areas affected by drought and flooding (IPCC, 2007a). Of particular concern is the vulnerability of mountain glaciers and snowpack and the risk of severe loss of water resources. A recent example is the 8-year drought in the Colorado River basin (1999–2007 water year), which is the most extreme in the measured hydrological record (100 years). Within the past decade, many communities in southern California have experienced their single driest year on record (CDWR, 2008a). Even the southeastern United States is in a drought (Box 2.4).

Some of the most alarming findings in the IPCC Fourth Assessment, such as the estimate that dryland areas have doubled since the 1970s (IPCC, 2007b, Chapter 3), are not based on direct

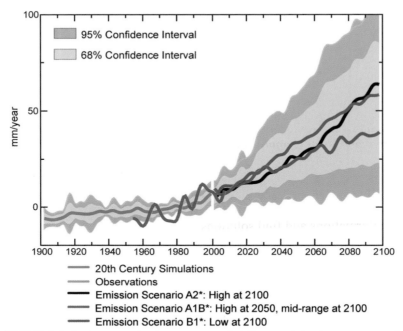

FIGURE 2.6 Increase in the amount of daily precipitation over North America that falls in heavy events (the top 5 percent of all precipitation events in a year) compared to the 1961–1990 average. Various emission scenarios are used for future projections. Data for this index at the continental scale are available only since 1950 (pink line). SOURCE: Karl et al. (2008).

measures such as soil moisture, but rather on crude estimates based on running averages of precipitation and air temperature. The large uncertainties highlight the need for future investments in observations, models, and process understanding. However, current climate information can be used to support decisions on water resources at a variety of geographic scales, even at the current skill levels of hydrological forecasts (Beller-Simms et al., 2008). A host of freshwater governance options have emerged, including mechanisms for making decisions under uncertainty and adaptive management to enable freshwater systems to respond to different kinds and magnitudes of impacts (Ivey et al., 2004; Olsson et al., 2004; Pahl-Wostl, 2007; Werick and Palmer, 2008).

In summary, the twentieth century has witnessed fundamental changes in the hydrological cycle from global to watershed scales. From a physical perspective, coupled ocean–atmosphere models are able to account for some of these changes, but poor understanding of the forcing terms (e.g., aerosols, land surface changes), coarse resolution of the models (few hundred kilometers), and lack of data over the oceans limits the ability of models to capture regional and local changes. From a socioeconomic perspective, there has been considerable research on water but much less that also considers water governance and climate. An integrated approach that takes account of physical, social, and ecological factors affecting freshwater change is needed to understand the potential transformations and find solutions.

BOX 2.4 Drought in the Southeast, a Wake-Up Call

Parts of the southeastern United States have been experiencing drought conditions since 2005 or early 2006 (Figure). The southeastern drought is instructive for two reasons: it provides an example of drought in a humid part of the United States, where water scarcity is not typically seen as a major challenge; and it illustrates the region's relative lack of preparedness for drought, especially compared to the arid West, where drought is a prominent water management concern. The Southeast has experienced significant population growth, but has not invested in the major interregional water infrastructure and institutional arrangements that might have allowed it to respond to drought. Although decreasing rainfall was well monitored by state climatologists, the impacts to agriculture, fisheries, and municipal water supplies may have been made worse because the involved states (Georgia, Alabama, and Florida) failed to act on the water resources compacts between them (Feldman, 2007, cited in Beller-Simms et al., 2008). The states could not agree on water allocation schemes and so let the compact expire. Faced with the tough decision of either relying on the forecasted above-average Atlantic hurricane season or being more conservative, the governor of Georgia instituted state-level outdoor water restrictions[a] and declared a state of emergency for parts of the state in October 2007. Georgia then filed a motion in federal district court for a preliminary injunction to require the U.S. Army Corps of Engineers to reduce releases for downstream water demands, including mandated flows for aquatic species in Florida listed under the Endangered Species Act. In addition, Georgia's Senate Resolution 822, introduced in 2008, called for establishment of Georgia-North Carolina and Georgia-Tennessee boundary-line commissions to survey and settle disputed state boundary locations that, if settled in Georgia's favor, would place portions of waterways such as the Tennessee River within Georgia. This case shows that even when climate information is available, unresolved conflicts between upstream and downstream user priorities constrain their use for mitigating negative impacts (Beller-Simms et al., 2008).

FIGURE Low water levels in Lake Lanier, the main source of drinking water for Atlanta, shown on November 16, 2007. SOURCE: Ed Jackson, University of Georgia.

[a] http://www.caes.uga.edu/topics/disasters/drought/.

Research Needs

Specific research needs include the following:

- Prediction of changes in water supply (runoff, groundwater, snowpack) and the reliability of the water supply, which requires improvements in decadal modeling, regional modeling, and understanding and modeling of the land surface hydrological sensitivity to climate change (Graham et al., 2007)
- Understanding the causes and predictability of extreme events (e.g., droughts, floods)
- Improve understanding and predictability of updated watershed-level rainfall-runoff relationships that account for increased precipitation intensity for flood forecasting purposes, especially for locations prone to rain-on-snow flood events

- Prediction of changes in water demand, which require demographic models that incorporate climate change impacts and models that consider the effects of climate change on natural and agricultural landscape water use
- Research on water governance, including adaptive management models, adaptive capacity building, and water systems sustainability
- Research on the economics of water supply, demand, and conservation; and on human perceptions and valuations of impacts, cost of adaptation, and equity, which are needed to inform adaptive action
- Development of long-term observations and tools for predicting hydrological variables of most value to water resource managers (e.g., timing of snowmelt, groundwater recharge rates) from climate model output

Better datasets are needed for determining decadal- to longer scale trends in regional forcing terms because of aerosols (specifically absorbing aerosols, which would be measured by the Aerosol-Cloud Ecosystems mission recommended by the National Research Council's Decadal Survey [NRC, 2007b]) and land-surface modification (specifically land cover change, which would be measured by the Landsat Data Continuity Mission). Precipitation measurements over land and the oceans are critical for both basic climate science and water resources applications and would be made by the Global Precipitation Mission, recently given high priority in the Decadal Survey. Global measurements of stream flow, soil moisture, and evaporation are also needed. For example, the Soil Moisture Active and Passive Mission, also recommended by the Decadal Survey, would provide data for drought monitoring and for driving predictive models of water balance. Socioeconomic data needs include water demand, consumption patterns, scarcity, equity, distribution, and adaptation costs. A comprehensive review of operational ground-based monitoring networks (e.g., Snowpack Telemetry Network) would reveal whether they are adequate to detect climate change impacts.

AGRICULTURE AND FOOD SECURITY

In 2007, an estimated 923 million people were seriously undernourished, 75 million more than in 2005.[6] Climate change is expected to alter the global food supply, with implications for global and regional agricultural production and food security. Indeed, the effects of climate change on food availability and the stability of the food system are already being felt, especially in rural locations where crops fail or yields decline, and in areas where supply chains are disrupted, market prices increase, and livelihoods are lost (FAO, 2008).

Some important agricultural areas appear to be experiencing significant deviations from the average climatic conditions under which the current farming systems developed, causing considerable hardship (e.g., Box 2.5). A record-setting severe winter in Central Asia in 2007 followed by large snowfall and severe flooding have threatened food security in the region, particularly in Tajikistan and Kyrgyzstan.[7] Afghanistan and Iraq are currently experiencing the worst drought in 10 years, adding to those nations' woes.[8] Such problems can also stress less vulnerable countries. The extreme heat wave of 2003 in France and Italy resulted in uninsured economic losses of EUR 13 billion for the agricultural sector (IPCC, 2007c, Chapter 5; see also Box 2.9).

Understanding and predicting regional climate trends and their impact on agriculture and the national food supply is a high priority for governments.[9] Climate change affects both commercial farming, which is often an integral part of national economies, and subsistence farming, which is common in developing countries and determines the livelihood of millions of people. This latter group is by far the most vulnerable to climate variability and change (Parry et al., 2005), and it is well recognized that poor, natural resource-dependent, rural households will bear a disproportionate burden of the adverse impacts of climate change (Mendelsohn et al., 2007; Agrawal, 2008).

---

[6] *http://www.fao.org/newsroom/en/news/2008/1000945/index.html.*
[7] *http://www.fao.org/giews/english/shortnews/casia080408.htm.*
[8] *http://www.pecad.fas.usda.gov/highlights/2008/09/mideast_cenasia_drought/.*
[9] *http://www.earthobservations.org/cop_ag_gams.shtml.*

BOX 2.5 Catastrophic Long-Running Agricultural Drought, Murray Darling Basin, Australia

The Murray Darling Basin provides 85 percent of the water used for irrigation in Australia and has traditionally produced 40 percent of the country's fruit, vegetables, and grain. Whereas drought typically occurs once every 20 years or so, over the past 7 years it has become an annual occurrence. The Australian government has spent about $2 billion dollars in the past few years in "exceptional circumstances" payments to help the affected farmers. It also offered $150,000 to farmers who decide to leave their land. Over the past 5 years, extreme drought conditions have forced more than 10,000 farmers off the land. Many ranchers have had to sell off their stock, the remaining farmers have had to use water more efficiently, and severe water restrictions have been introduced in urban areas in the region. The competing demands for water for domestic use, irrigation, and ecosystem preservation far exceed the recent flow of the three main rivers. A U.S. $3.6 billion emergency water conservation plan is being put in place, but it is uncertain whether this plan will be enough if the drought continues or if it will be implemented in time to make a significant difference.

FIGURE Dead trees and cracked earth on a farm near Kerang, a district in the Murray Darling Basin about 360 km north of Melbourne, August 24, 2007. SOURCE: REUTERS/Tim Wimborne.

Australia can experience strong ENSO events with devastating events on rangelands and agriculture. Severe droughts have occurred throughout the country's history, for example in 1900, 1942, 1982, and 1992. The IPCC Fourth Assessment projects that there will be up to 20 percent more droughts in the region by 2030 and a decrease in annual Murray Darling River flow by 10 to 25 percent by 2050.

The La Niña conditions in 2006 failed to bring its usual heavy rains to Australia, highlighting the need for improved seasonal to interannual regional climate forecasting to better predict rainfall and temperature over the next season and likely trends over the next few years. A better understanding of climate trends would place federal and regional governments in a better position to manage the resulting economic impacts and population displacement, to help their most vulnerable citizens, and to promote effective adaptation strategies.

SOURCES:
http://news.bbc.co.uk/2/hi/asia-pacific/7499036.stm,
http://www.independent.co.uk/news/world/australasia/australias-epic-drought-the-situation-is-grim-445450.html,
http://www.bom.gov.au/climate/drought/livedrought.shtml,
http://www.environment.gov.au/water/mdb/index.html,
http://news.nationalgeographic.com/news/2007/11/071108-australia-drought_2.html.

Tropical crop production is likely to suffer under a warming climate, whereas mid- to high-latitude regions could benefit initially from a small amount of warming (IPCC, 2001a). In its Fourth Assessment report, the IPCC noted that the large majority of climate models predict a decrease in precipitation in the subtropics by the end of the century and an increase in precipitation extremes in southern and eastern Asia, east Australia, and northern Europe (IPCC, 2007c, Chapter 5). Declines in water availability are projected for the Mediterranean Basin, Central America, subtropical Africa, and Australia. The Southwest and mid-continental agricultural areas of the United States are also expected to have droughts, reducing crop production and/or increasing demands for water in an area that is already beginning to experience water conflicts. The last U.S. national assessment (Reilly et al., 2000) concluded that the net effect of the climate scenarios studied on the agricultural sector over the twenty-first century is generally positive. A more recent and detailed assessment of the impacts of climate change on U.S. agriculture, crops, rangelands, and livestock (Hatfield et al., 2008) presents a less optimistic picture.

Although the Food and Agriculture Organization projects about a 60 percent decrease in the growth rate of food production, an 80 percent increase in agricultural production by 2050 is required to feed a growing population (FAO, 2008). This need, in turn, will require new croplands to be cultivated, many of which will replace tropical woodlands and forests in sub-Saharan Africa and Latin America (Ramankutty et al., 2002). Further increases in cultivated area may be needed if an increased frequency of climate extremes lowers yields below projections (Easterling et al., 2007). Shrinkage of mountain glaciers will decrease the water available, but increasing evaporation will increase the need for irrigation.

The stress of climate variability and change on the global food supply will be exacerbated by population growth, increased wealth in developing countries, rising cost of fertilizer, political instability, national policies, and pests and invasive species. For example, wheat fields in the United States are now being planted with corn for ethanol, driven by the increased cost of gasoline, the demand for alternative energy sources (in this case biofuel), and government subsidies (USDA, 2007). Crop failures around the world in 2007 led several countries to meet national needs by restricting crop exports, thus reducing global supply (Trostle, 2008).

Research Needs

Basic and applied research, supported by modeling, field studies, and satellite observations, are needed to provide an improved understanding of global agricultural land use, productivity, and food supply in the context of a changing climate. The research needs fall into two broad categories: climate modeling for agriculture and global and integrated modeling of agricultural land use and associated mitigation and adaptation options.

Climate Modeling for Agriculture

Current predictions are largely inadequate for food security systems. Improved climate models are needed (1) that generate output at regional scales with improved timeliness and skill (Mukhala and Chavula, 2007) targeted for specific agricultural needs (Meinke and Stone, 2005), and (2) that include inputs to key processes in crop models related to climate change (i.e., temperature, water stress, and their interaction with elevated CO_2; Tubiello and Fischer, 2007; Tubiello et al., 2007). Matching the spatial and temporal scales of climate and crop models is a necessary integrative step (Challinor et al., 2007). Adjustments in rain-fed and irrigated agriculture depend on projections of water supply and demand over the next few decades, particularly for vulnerable semi-arid areas (see Box 2.5 and the "Freshwater Availability" section). These projections should quantify the rates of glacial retreat and changes in precipitation for mountain systems which feed irrigated lands (e.g., in Central Asia). Recent efforts to enhance drought monitoring in the United States will need to be replicated for drought-prone regions of the world with populations and liveli-hoods at risk.

Global and Regional Integrated Monitoring and Modeling of Agri-cultural Land Use and Associated Mitigation and Adaptation Options

A new generation of integrated dynamic Earth system models that incorporate both physical and socioeconomic factors is needed to better project changes in regional food supply and demand re-sulting from a changing climate and to inform mitigation and adaptation options (Howden et al., 2007; Ingram et al., 2008). Ef-fective adaptation will require an integrated view of climate change issues, including climate variability and market risk in the context of regional economic and sustainable development (Adger et al., 2007). It will require effective institutions for determining agricultural production and implementing adaptation measures at a range of scales. Tariffs and subsidies that strongly influence the global supply of food will inevitably change as governments re-spond to shifting markets and changes in global agricultural supply and demand, resulting in part from changes in regional climate (Tubiello and Fischer, 2007). At the local scale, institutions and institutional partnerships will have to be strengthened to increase access to adaptation methods (Agrawal, 2008).

Regional, spatially explicit, process models of land-use change are needed to project agricultural expansion, intensification, and abandonment and to model the potential impacts of these changes on the major biogeochemical cycles, land–atmosphere exchange of water and energy, and human population dynamics. Place-based models should explore societal vulnerability and the various autonomous and planned adaptation pathways and coping strategies. Climate change at the low end of the anticipated range over the next few decades is likely to have only modest economic impacts on U.S. agriculture (Mendelsohn et al., 1994; Kelly et al., 2005; Schlenker et al., 2005), but the impacts in other areas of the world could be larger. Similarly, the economics of adaptation in agriculture is poorly understood.

End to end, coupled biogeochemical, hydrological, and economic models should address the impacts, feedbacks, and costs of different agricultural land use (e.g., extensification, abandonment) and different land-use mitigation options to sequester carbon (e.g., conservation agriculture, no-till agriculture, shade cropping), to establish the impacts of agricultural intensification and increasing fertilizer use (e.g., see Box 2.7). Similarly, improved scientific understanding is needed to examine the trade-off between using crops for food or for biofuel and the impacts on food prices, secondary land use, and soil erosion. Field experiments will be needed to parameterize these models and to quantify the net carbon sequestration and the water-use and -quality implications associated with different mitigation and alternative energy options (e.g., NRC, 2008d).

Priority Infrastructure Needs

The infrastructure needed to support the aforementioned science includes increased computational capacity to run higher resolution climate models with regional specificity (see "Earth System Modeling" in Chapter 3) and improved land surface observations.[10] Continuous satellite measurements, such as following the Moderate Resolution Imaging Spectroradiometer with the Visible Infrared Imaging Radiometer Suite, are needed for monitoring agriculture. Long-term moderate resolution (i.e., Landsat class) observations will be needed, but with an increased temporal frequency (i.e., 3- to 5-day coverage) to monitor changes in cropland and crop area and to drive crop production models and famine

---

[10] A summary of the observational infrastructure needed for agricultural monitoring can be found at *http://www.earthobservations.org/cop_ag_gams.shtml.*

early warning systems.[11] The inadequacy of U.S. spaceborne observations in this respect has led the U.S. Department of Agriculture to become the single largest purchaser of Indian satellite data, which are now used for monitoring U.S. crops. Targeted high-resolution (1 to 3 m) imaging is also needed to monitor crop conditions in subsistence agricultural regions, to improve national agricultural production estimates, and to help monitor the agricultural aspects of carbon management. Recent advances in microwave remote sensing for agricultural monitoring also warrant further investigation.

MANAGING ECOSYSTEMS

Humans actively manage ecosystems to provide food, water, timber, and other resources. We rely on ecosystems to regulate local climate conditions and remove pollutants from the air and water (Millennium Ecosystem Assessment, 2005). Although many of these services are already under stress due to pollution, overuse, land-use change, and other anthropogenic factors, climate change will further affect the ability of ecosystems to sustain these services and natural resources (IPCC, 2007c, Chapter 4). Consequently, it will be important to understand the linkages between ecosystems, societies, and climate; to assess human and ecosystem vulnerabilities to climate change; and to devise management strategies that mitigate climate change (e.g., decreasing deforestation rates or planting forests to sequester carbon) while preserving ecosystems and their services.

The current distribution of plant and animal species over large areas of the globe reflects human appropriation of primary production (Haberl et al., 2007), alteration and fragmentation of habitat, and modifications of the energy, nutrient, and water cycles. Modern ecosystems are also responding to observed climate changes, as recorded by changes in the timing of phenological events (e.g., leaf-out, flowering), migration patterns, and the ranges of fish and marine mammals (IPCC, 2007c). The rapid rate of climate change, combined with human-induced stressors (e.g., poor land manage-

---

[11] *http://www.earthobservations.org/documents/cop/ag_gams/20070716_ geo_igol_ag_workshop_report.pdf.*

ment practices), may outpace the ability of ecosystems to adapt, leading to steep declines in biodiversity and ecosystem resilience (IPCC, 2007c). This issue was recognized in Article 2 of the United Nations Framework Convention on Climate Change, which states that stabilization of greenhouse gases "should be achieved within a timeframe sufficient to allow ecosystems to adapt naturally to climate change." The documentation of large-scale mortality events, such as those observed for pinyon-pine forests in the southwestern United States (Breshears et al., 2005) or rapid declines of coral reefs (Box 2.6) suggests that this goal is not being achieved.

BOX 2.6 The End of Coral Reefs?

Coral reefs provide critical habitat to support fisheries and marine biodiversity. They benefit humans by supporting fishing and tourism, supplying natural products, and forming a breakwater that helps protect coastal property from storm and wave damage. The global net economic benefit of reefs has been estimated at $30 billion per year (Cesar et al., 2003), including several billion dollars per year in Florida and Hawaii (Johns et al., 2001; Cesar et al., 2002).

The past several decades has seen a dramatic increase in coral mortality and reef degradation (Pandolfi et al., 2003). A third of reef species are currently in danger of extinction (Carpenter et al., 2008). The reasons for this decline include coastal development, increased disease, overfishing, pollution, and climate change (Buddemeier et al., 2004; see also figure). Climate change affects coral reefs in several ways. Elevated atmospheric CO_2 decreases ocean pH and carbonate ion content, reducing the ability of corals to form the calcium carbonate skeletons that form the reef structures, and ultimately undermining reef structures and their ability to support biodiversity (Kleypas et al., 2006). Warmer ocean temperatures may increase the geographic range that coral reefs can develop, but also increases coral bleaching, caused when the coral expels its algal symbiont. Sea level rise, which changes the intensity of coastal storms and coastal erosion regimes, will also negatively affect coral reef structures. In heavily populated regions where most coral decline is observed, existing stresses weaken the ability of coral reefs to adapt to climate change. However, even corals in "pristine" areas are affected by climate change and elevated CO_2 levels.

No analogs of ocean chemistry exist in the historic record to help us predict the long-term response of coral reefs to climate change. However, coral reefs are already in steep decline, creating the potential for tremendous consequences for marine ecology, biodiversity, and local economies. Research is needed on the basic biology of coral reefs, especially the causes of observed declines and the effects of elevated water temperature and CO_2. Human behaviors driving changes in reef ecosystems, as well as consequences of the loss or alteration of coral reef ecosystem services have to be understood to design effective management strategies.

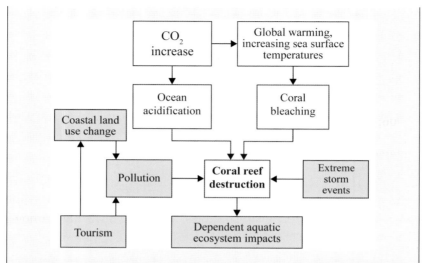

FIGURE The interactions of multiple stressors related to coral reef decline. Stressors that are direct results of climate change are in white boxes; stressors that are related to human use of corals or that may be altered by climate change (e.g., extreme weather events) are in blue boxes. Note that there are feedbacks between coral reef decline and some human activities (e.g., tourism), which in turn affect reefs and how they are managed.

Climate change will affect ecosystems through a number of mechanisms, such as by altering patterns of temperature, rainfall, ocean stratification, upwelling, and mortality rates caused by extreme events, such as storms, fires, and coastal hypoxia (Box 2.7). Interactions among individual organisms responding to climate changes are complex and may lead to threshold responses (or tipping points) with rapid changes in ecosystem productivity, composition, or location. For example, warmer winter temperatures have been linked to increased prevalence and intensity of shellfish diseases (Cook et al., 1998; Soniat et al., 2008), and earlier spring blooms of marine plankton which affect fishery production (Harrison et al., 2005). Interactions between temperature and metabolic demand by marine organisms, coupled with the effects of temperature on oxygen solubility, can lead to nonlinear responses in fish productivity (Del Toro-Silva et al., 2008). Climate-mediated physiological stress resulting from warming temperatures can also compromise disease resistance of marine and terrestrial organisms and result in emergence of new diseases, increased frequency of opportunistic diseases, and exposure of previously uninfected host populations to pathogens (e.g., Harvell et al., 1999, 2002, 2004).

Even when climate change is minimal, ecosystems will respond to elevated CO_2 levels in the atmosphere, through either altered photosynthesis on land (Norby et al., 2005) or reduced calcification rates in the oceans (Orr et al., 2005; Kleypas et al., 2006; Box 2.6). The effects of these changes on individual organisms and ecosystems are only beginning to be studied, and their interactions with climate responses are not well known. For example, increased productivity in forests from elevated CO_2 may add resilience of forests to storm or insect damage (Negrón et al., 2008). Excess CO_2 absorbed into the ocean may increase calcification by some phytoplankton species, thereby enhancing carbon export to the deep ocean (Iglesias-Rodriguez et al., 2008), but cause irreparable damage to other organisms (Box 2.6; Orr et al., 2005). Other effects of atmospheric composition on marine and terrestrial ecosystems include altered sunlight for photosynthesis with increased aerosol loading and effects of pollutants (especially nitrogen deposition and ozone) on productivity.

In many locations, responses to climate or atmospheric changes will add to other human-related stressors such as pollution, land use, and the introduction of invasive species to affect ecosystem resilience. For example, increases in storm runoff and warmer, more stratified coastal ocean conditions may exacerbate "dead zones" (Box 2.7).

Ecosystem responses to climate change or large-scale management will in turn modify climate on a variety of spatial scales. Land and ocean ecosystems are a key component of the climate system. They process energy, water, carbon, and nutrients, and mediate the fate of incoming sunlight, which in turn influence factors such as cloud formation and greenhouse gas fluxes. Some of these mechanisms may amplify or dampen climate change through altered surface energy balance and/or greenhouse gas emissions.

Management strategies that involve manipulation of ecosystems to mitigate climate change (e.g., sequestering carbon by planting trees, managing forests, or fertilizing the oceans) will involve a spectrum of human–climate–environment interactions. In evaluating such strategies, not only the target impact (e.g., sequestration) but

BOX 2.7 Enter the Dead Zone

When excessive amounts of nutrients (usually from fertilizer) flow into coastal waters, massive amounts of organic matter (e.g., algae) are produced, which consume oxygen as it decays and thus create "dead" zones. Dead zones in coastal oceans have expanded exponentially since the 1960s, with the number doubling each decade. There are currently 400 dead zones, covering a total area of more than 245,000 km^2 (Diaz and Rosenberg, 2008). The largest in the United States is at the mouth of the Mississippi River (Figure), which carries nitrogen and phosphorus from chemical fertilizers used in agriculture. In 2008 this dead zone reached its second largest extent (21,000 km^2, approximately the size of New Jersey), fed by high nitrate loads (37 percent higher than 2007 and the highest recorded since measurements began in 1970) from the Mississippi and Atchafalaya rivers. Its extent would have been greater if not for aeration of the waters caused by the passage of Hurricane Dolly.[a]

FIGURE Mississippi River plume (brown water at left) meets the Gulf of Mexico (blue water at right) at Southwest Pass. SOURCE: N. Rabalais, Louisiana Universities Marine Consortium. Used with permission.

Hypoxia (very low levels of dissolved oxygen) is one of multiple stressors affecting aquatic ecosystems (Figure, below); others include overfishing, habitat loss, toxic algal blooms, and climate change. Although not the direct cause of coastal hypoxia, climate change affects environmental conditions that can affect the extent and/or likelihood of dead zones. For example, projected increases in rainfall will increase river discharge, nutrient delivery, and stratification (via freshwater influx) of the upper water column, thereby possibly expanding coastal regions impacted by dead zones (Justic et al., 1997; Rabalais et al., 2002; Diaz and Rosenberg, 2008). On the other hand, more frequent storms may mitigate the development of dead zones. Changing circulation patterns may contribute to formation of regions of coastal hypoxia. The dead zone reported off the coast of Oregon in 2006, which extended 3000 km, is an example of the contribution of atmospheric and coastal circulation to the development of a dead zone. Wind patterns over this region intensified upwelling of shelf water with low dissolved oxygen, exacerbating the existing biologically-produced hypoxic conditions in the area (Chan et al., 2008). Wind-driven coastal upwelling is a response to large-scale atmospheric and oceanic circulation patterns, which in turn are influenced by global warming. The Oregon hypoxia event is troubling because eastern boundary current systems such as this one are among the most productive marine ecosystems in the world.

Warmer waters resulting from a warming climate will dissolve less oxygen, which may enhance low-oxygen regions. An ocean biogeochemical model driven by the "business as usual" emissions scenario predicts reduced oxygen levels and increased extent of oxygen minimum zones in the oceans in the next two centuries (Schmittner et al., 2008). Ocean temperatures will also affect O_2 metabolic demand by fish (Del Toro-Silva et al., 2008), which may increase fish mortality rates and expand dead zones. These examples illustrate how climate change may become more of a critical contributor to low-oxygen regions. Understanding the occurrence and extent of dead zones requires research on the complex interactions between land use, coastal ecosystems, environmental conditions, and climate change. Such knowledge is needed to develop strategies to minimize and mitigate the effect of these events.

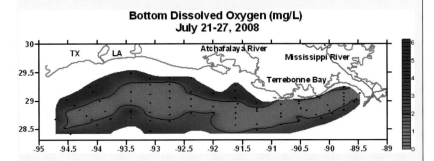

Bottom Dissolved Oxygen (mg/L)
July 21-27, 2008

FIGURE Oxygen levels (in ppm) in the bottom waters of the Gulf of Mexico dead zone, from July 21 to 27, 2008. The red area enclosed by the black line is in a state of hypoxia. SOURCE: N. Rabalais, Louisiana Universities Marine Consortium. Used with permission.

[a] *http://www.gulfhypoxia.net/research/shelfwidecruises/2008/Press Release08.pdf.*

also the consequences for ecosystems, non-CO_2 climate effects (e.g., heat and water budgets, other greenhouse gases), the vulnerability of carbon storage over longer timescales, and economic trade-offs (e.g., value of sequestration versus avoided emissions) need to be considered (Dilling et al., 2003). For example, private corporations are making plans for large-scale releases of iron to the oceans to generate carbon offsets, despite scientific uncertainty about the efficacy and timescale of iron fertilization for carbon sequestration and the ecological consequences of such additions (Buesseler et al., 2008). Strategies such as afforestation not only change carbon budgets but also affect water fluxes and stream levels across the United States (Jackson et al., 2005). Tree planting programs in boreal and tropical ecosystems will not have the same net climate effect because of the disproportionate effect of boreal trees on reflectivity of the land surface (Bala et al., 2007). Forests planted to sequester carbon will be vulnerable to extreme weather events such as hurricanes that can cause large-scale mortality (Chambers et al., 2007). Feedbacks between drought, forest mortality, fire, and land clearing in tropical forests may lead to a critical loss of biodiversity and to a shift to overall drier climates in tropical regions (Bonan, 2008). There will inevitably be trade-offs between use of land for climate mitigation and for other priorities. For example, a broad mitigation strategy for sustained reduction of emissions from deforestation and degradation will have to consider the livelihoods of the people living in the forested regions (Malhi et al., 2008). Despite the sustainable development requirement embedded in the Kyoto clean development mechanism, ecosystem management programs have fallen considerably short of their original goals (Bozmoski et al., 2008).

All lands and coastal areas in the United States are managed, even if management takes the form of a decision to leave lands "wild." However, changes in climate, atmospheric composition, and pollutant deposition will affect even set-aside areas. Considering climate change in decisions on federal land management poses both scientific and regulatory challenges (Julius et al., 2008), including the ability of current regulatory frameworks such as the Endangered Species Act (ESA) to incorporate projected climate change impacts into permitting decisions (see Box 2.8).

BOX 2.8 Climate Change and the Endangered Species Act

FIGURE The polar bear was listed as threatened by the U.S. Fish and Wildlife Service in May 2008 and a Special Rule under the Endangered Species Act providing for its conservation was issued in December 2008. SOURCE: Dave Olsen, U.S. Fish and Wildlife Service.

Concern over the future of the polar bear—the first species listed as threatened due to climate concerns—centers on loss of habitat associated with warming of the Arctic and sea ice melting. However, climate change will clearly affect the ranges of many individual species, the future suitability of areas set aside to preserve habitat, and plans for managing discrete populations of species of concern. As presently authorized, the ESA is focused on single-species management rather than ecosystem management, an approach that does not readily facilitate adaptation. Reauthorization of the ESA to address this and other aspects of how the federal government manages species of concern has been a subject of proposed legislation and much commentary over the past decade.[a] Experience to date with ESA administration and compliance has demonstrated the Act's strong influence on land and water management decisions and the corresponding costs of recovery plan implementation for the regulated community, even absent the further complications associated with anticipated climate change. Options for ecosystem adaptation could include changes to the existing regulatory framework to incorporate climate change into species recovery planning and to focus funds available for recovery plan implementation on strategies that provide resiliency. Policy decisions such as this—how to amend current legislation or to create a new legal or administrative regulatory framework for including climate change in key legislation—will frame needs for future scientific research on key species and ecosystems and their responses to climate change.

---

[a] See, for example, *http://www.publicland.org/endangerSpecies.htm*.

Research Needs

An enhanced program of basic and applied research is needed to improve understanding of the responses of marine and terrestrial ecosystems to climate change, the major impacts to and vulnerabilities of ecosystems and the services they provide, and the role of human actions and nonclimate stressors in facilitating or ameliorating the potential for rapid ecosystem responses to climate change. CCSP Synthesis and Assessment Product 4.4 (Julius et al., 2008) discusses the design of climate change adaptation strategies for a subset of federally managed lands that balance resource needs with preservation of biodiversity and key ecosystem services under a changing climate, as well as the evaluation of the societal costs of specific management actions (or inaction). Research is also needed to assess ecosystem and land management options that can help in preserving biodiversity and sequestering carbon under a changing climate and to evaluate the societal costs of specific management actions (or inaction). Specific research needs include the following:

• Assessment of key vulnerabilities of ecosystems to climate and other human stresses, including the compound effects of multiple stresses and the potential impacts of extreme or abrupt events (e.g., heat waves, extended drought, increased severe weather or flooding, changes in sea ice extent, changes in ocean circulation).

• Mechanisms and timescales for adaptation of ecosystems and ecosystem management to climate and other changes, and the possibility of thresholds leading to, for example, ecosystem collapse, extinctions, or regional-scale mortality events. Long-term datasets are required to investigate paleovegetation shifts and rates of ecosystem migration. New research is needed to separate the effects of climate from other changes, such as those caused by invasive species, pollution, land-use change, landscape fragmentation, or the loss of high-level predators.

• The net impact of increased CO_2 levels, including ocean acidification, in combination with other stressors (pollutants, nutrient deposition) on ecosystems, especially for important ecosystems such as tropical forests and coral reefs.

• The consequences of changing ecosystems on climate feedbacks and human vulnerabilities. This requires assessment of the full suite of climate feedbacks, including carbon, surface energy and water balance, aerosols, and clouds. Key uncertainties are associated with areas vulnerable to rapid change in land cover, including the Arctic and tropical regions.

- The human behaviors associated with natural resource use, such as harvesting and land-use change, that affect trophic structures and change ecosystems, and how these behaviors change in response to ecosystem stress or change. The effects of management strategies on climate, ecosystem services, and the resilience of ecosystems to climate change will need to be assessed. Field experiments and models can be designed to learn about coupled human- and environmental systems and to test different management interventions (i.e., adaptive management).
- The valuation of ecosystem services, including the economic and other costs associated with impacts of climate and other environmental changes.
- How managed ecosystems function, including those associated with growing urban areas, and how to improve the provision of ecosystem services in human-dominated landscapes. We need to understand how to manage the trade-offs between services with direct contributions to local human livelihoods and those with more indirect or global contributions, such as carbon sequestration and biodiversity preservation.
- Adaptive approaches and institutional and governance mechanisms for addressing the regulatory aspects of special status species management.

To make significant progress, the research will have to be supported by observation networks to document ecosystem changes over time and across types of ecosystems and human interactions through continuous measurement of variables such as greenness, land cover, and ocean color. An ocean observatory network (ORION Executive Steering Committee, 2005) and a continental-scale land observatory network (Keller et al., 2008) have been planned and are moving into implementation, although it may be years before operational systems are created that can be used to forecast environmental changes and their effects on biodiversity, coastal ecosystems, and climate. Improved models with better process understanding are needed (1) to evaluate the combined impacts of climate and other environmental stressors on a range of ecosystems at regional scales, and (2) to integrate and evaluate human drivers of and responses to environmental change in dynamic feedbacks with ecosystem models. Finally, tools are needed to inform adaptive management strategies and estimate the resilience of various ecosystems under scenarios of climate, nutrient, water, and human systems change.

HUMAN HEALTH

Climate change has been called the greatest regressive tax in history, with the populations imposing the least stress through their greenhouse gas emissions experiencing the greatest health impact and vice versa (Figure 2.7). This relationship exists because, for the most part, climate change does not create new diseases and other health risks, but exacerbates existing ones. Poor health not only amplifies vulnerability, but also reduces the ability of communities and individuals to cope with or adapt to climate and other stresses (IPCC, 2007c, Chapter 8). Substantial inequalities in coping capacity exist, and perhaps are growing, worldwide, including in the United States, where poor, elderly, uninsured, and minority populations are much more vulnerable. In the United States, a robust public health infrastructure, such as sanitation and wastewater treatment facilities, has proven the best defense against adverse health effects from climate change (Gamble, 2008).

The most authoritative assessment of the impacts of climate change on health to date was done in conjunction with the Comparative Risk Assessment (CRA) project of the World Health Organization (Ezzati et al., 2004; McMichael et al., 2004). It found that for only five outcomes (malnutrition, diarrhea, malaria, flood injuries, and cardiovascular disease), 160,000 premature deaths annually could be attributed to climate change in 2000, or 0.4 percent of the global burden of disease. Some 88 percent of the attributable burden fell on poor children because of their existing vulnerability to diarrheal diseases, malnutrition, and malaria. Climate change impacts in the second CRA assessment are expected to be significantly larger, even though the base year will only be 5 years later (2005).

FIGURE 2.7 Cartogram of climate-related mortality from malaria, malnutrition, diarrhea, and inland flooding per million people in the year 2000. The sizes of the regions are proportional to the increased mortality. SOURCE: Patz et al. (2007), Figure 1b. Reproduced with kind permission from Springer Science and Business Media.

Climate change health impacts are divided into five categories:

1. Direct impacts through changing weather patterns (e.g., storms, floods, temperature extremes)
2. Indirect impacts through changes in water supply, water quality, and air pollution, and in ecosystems leading to shifts in disease vectors
3. Systemic impacts through shifts in food supplies, refugee patterns, coastal and agricultural livelihoods, and society's responses to climate change, such as geoengineering, carbon taxes, and biofuel production
4. Low-probability high-consequence impacts, such as extremely rapid climate change or sea level rise
5. Cobenefit impacts (sometime called "no regrets" strategies), in which climate mitigation efforts are chosen to help protect health by reducing health-damaging air pollution emissions, lowering the vulnerability of poor populations, improving the built environment, and other means

In general, the ability to quantify the size and distribution of health impacts using standard biomedical tools such as validated exposure models and epidemiology is highest for category 1 and declines for categories 2 and 3. Category 3 impacts may be the most important for health over the long run. Few attempts have been made to quantify effects in category 4. The fifth and more positive type of impact is also of substantial research and policy interest.

Direct Health Threats

Changes in weather and storm patterns will engender many of the most serious climate change threats to the health of the U.S. population. Heat waves are expected to become more intense, more frequent, and to last longer (IPCC, 2007b, Chapter 10). High temperatures and humidity can cause death or chronic illness from the after-effects of heat stress. Outdoor workers in the construction, agriculture, forestry, and fishing industries appear to be at particular risk. A heat wave in the midwestern United States led to approximately 600 heat-related deaths in Chicago over a period of 5 days in July 1995. Tens of thousands died in the 2003 heat wave in Europe (Box 2.9). These events have taught us much about human vulnerability and highlighted the need both for factoring climate information into public health systems and for integrating climate, social, and health research.

Climate change is expected to increase the risk of intense precipitation events and flooding (IPCC, 2007b, Chapter 10). Floods can overwhelm preparatory and coping systems, even in regions with long experience in flooding, as the 2008 floods in the midwestern United States revealed. In 2003, 130 million people were affected by floods in China alone. Rarely included in such data are the secondary deaths that follow from the unsafe and unsanitary conditions in the wake of floods.

Hurricanes are likely to become more intense with climate change, with impacts over broader scales (IPCC, 2007b, Summary for Policy Makers). Water supplies were contaminated with oil, pesticides, and hazardous wastes in the aftermath of Hurricane Katrina in 2005 (Manuel, 2006). Contamination of water supplies with fecal bacteria has led to diarrheal illness and some deaths following several hurricanes. As with heat waves, richer and poorer communities have different vulnerabilities (Adger et al., 2005). The insurance and reinsurance industry will be challenged to adequately cover the risk of such major events.

BOX 2.9 2003 European Heat Wave

In August 2003, a heat wave raised temperatures to more than 40°C (104°F) in France for 7 days (Figure), causing more than 14,800 deaths. Other European countries, including Belgium, the Czech Republic, Germany, Italy, Portugal, and Spain, also reported excess heat-related deaths, with total deaths of approximately 35,000 (IPCC, 2007c, Chapter 8). Older people were especially vulnerable, with about 60 percent of the deaths in France occurring in persons 75 years of age and older (Vandentorren and Empereur-Bissonnet, 2005). The lack of air-conditioning, inexperience coping with very high temperatures (e.g., need for hydration), and absence of nearby relatives were contributing factors. The extreme heat also caused other harmful exposures, such as increased tropospheric ozone and particulate matter. A French parliamentary inquiry found that existing systems for surveillance of heat wave stress were inadequate as were deficiencies in public health systems. Since that event, European governments have improved risk management systems, including better health warning and care of the elderly. Such heat waves are likely to increase with climate change and to further stress public health systems in a number of countries.

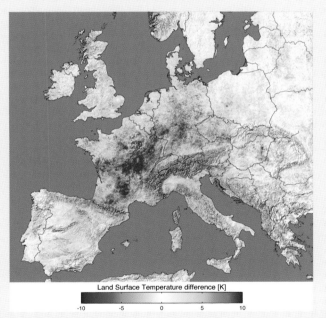

Land Surface Temperature difference [K]

-10   -5   0   5   10

FIGURE Differences in daytime land surface temperatures in Europe in 2003 from temperatures measured by the Moderate Resolution Imaging Spectroradiometer in 2000, 2001, 2002, and 2004. SOURCE: Image by Reto Stöckli, Robert Simmon, and David Herring, NASA Earth Observatory. Available at *http://www.iac.ethz.ch/staff/ stockli/europe2003/*.

SOURCES: Lagadec (2004), IPCC (2007c).

Droughts lead to regional water scarcity, salinization, disruption of food systems, and increased plant infectious diseases or pests, and affects human health through malnutrition, infectious diseases, and respiratory diseases. In the United States, the main health risks from drought depend on (1) the effects of temperature on the incidence of diarrheal disease (IPCC, 2007c); (2) linkages between water availability, household access to improved water, and the health burden associated with a range of diseases; and (3) the effects of temperature and runoff on the microbiological and chemical composition of water supplies.

Indirect Health Threats

Indirect effects of climate on health are linked with changes in ecosystems and include air quality, allergens, and vectorborne infectious and parasitic diseases. Concentrations of ground-level ozone associated with climate change are increasing in many regions and are implicated in pneumonia, asthma, other respiratory diseases, and premature mortality. Concentrations of other air pollutants, particularly fine particulate matter, may increase in response to climate change. Climate change may increase the frequency and severity of fire events, releasing toxic gaseous and particulate air pollutants, and possibly increasing the long-range transport of air pollutants such as aerosols, carbon monoxide, ozone, mold spores, and pesticides. Climate change has already caused an earlier onset of the spring pollen season in the northern hemisphere. Changes in the spatial distribution of natural vegetation may favor the growth of invasive plant species that cause allergies, such as ragweed.

Shifts in the natural reservoirs of disease vectors, such as mosquitoes, rodents, marine algae, birds, and deer, are already resulting in greater human exposures in some parts of the world, perhaps including the United States. Diseases such as malaria may return, although careful maintenance of the U.S. public health infrastructure may prevent them from becoming significant (Gamble, 2008). Dengue fever and Lyme disease may also rise in the country due partly to climate change. Climate change can shift the distribution of tick and mosquito vectors of disease (IPCC, 2007c, Chapter 8).

Diseases transmitted by rodents may also increase during heavy rainfall and flooding events.

Systemic Health Impacts

Although not well suited to standard assessment methods, systemic impacts are the most worrisome because of their potentially pervasive impacts on health. Changes in agricultural production due to climate change will likely lead to an increase in malnutrition, which is a major health risk globally. Malnutrition was the largest category of ill-health related to climate change in the 2004 CRA. The linkage is not straightforward, however, as malnutrition is also influenced by economic, social, and governmental factors, which vary in time and space. Similarly, shifts in refugee and other migration patterns due to sea level rise, persistent droughts, changes in agriculture, and other climate-related stressors are significantly mitigated or enhanced by specific social, economic, political, security, and geographic circumstances. Migrants and refugees exhibit substantially different health patterns and needs for public health services and medical care than do stable populations. The sustained ability to earn a livelihood is an important determinant of family health in all societies.

Health may also be affected by efforts to mitigate or adapt to climate change. For example, although carbon taxes may reduce emissions, they also can create "energy poverty" in the developed world, in which poor people are not able to afford to heat or cool their residences, and drive households in poor countries back to polluting solid fuels. Burning solid fuels is already responsible for 1.6 million premature deaths annually, twice as many as all urban outdoor air pollution. Biofuel subsidies, which can contribute indirectly to malnutrition, and geoengineering schemes such as injecting aerosols into the atmosphere are likely to have widespread impacts on health. Reliable assessments of these impacts, however, require cooperation and development of methods and databases across a number of disciplines.

Low-Probability High-Consequence Impacts

Several threshold changes might be triggered by climate-changing pollutants, including runaway methane emissions from the ocean floor or tundra, rapid melting of large ice sheets, and shifts in major ocean currents. Given the potential speed and magnitude of their impacts on climate and sea level, the resultant health impacts and those of society's responses could be large. Systematic assessments would require adapting probabilistic risk assessment methods used in other realms to these large complex systems.

Cobenefits Impacts

Given the substantial resources that may be required for climate mitigation, research aimed at understanding the impacts of reducing both greenhouse and other health-damaging pollution would be beneficial. Examples of research topics that may lead to measurable and cost-effective cobenefits for human health include:

- The effects of improved combustion methods on air quality and health
- The effects of different modes of transportation (e.g., walking, public transportation, driving) on air pollution, traffic, and obesity risks
- Ways in which the use of energy efficient materials and design affect household (and urban) environmental risks and energy demand
- The relationship between food prices, greenhouse gas emissions, and the health impacts of dietary choices
- The effects of alternative land use practices (e.g., reforestation, cultivation of biofuel plants) on human welfare and disease
- How changes in the use of contraception may affect health, population, and resource consumption

Showing which greenhouse mitigation efforts can yield short-term health and other benefits, even if they are intended primarily for protection from climate changes decades in the future, would improve the attractiveness and political viability of these investments.

Research Needs

Systematic assessments of current and future health risks from climate change are needed to help understand the total impact of climate change and thus to guide mitigation and adaptation efforts. Of particular need is a more complete understanding of the uneven pattern of health risks, both within and between populations, which are expected to have highly unequal impacts. High-priority research and health impact assessment activities include:

- The readiness of the nation to predict and avoid public and occupational health problems caused by heat waves and severe storms
- The potential U.S. health threats from changes in the pattern of disease vectors, such as birds, rodents, and mosquitoes under different scenarios of climate-induced ecosystem change
- Characterization and quantification of relationships between climate variability (trends or fluctuations in temperature, precipitation, or other weather parameters), health outcomes, and the main determinants of vulnerability and equity within and between populations (location, age, health status, etc.)
- Development of reliable methods to connect climate-related changes in food systems and water supplies to health under different conditions
- Estimation of disease burdens in all the main categories (direct, indirect, systemic) attributable to current climate change, from the global to the subnational level
- Prediction of future risks in response to climate change scenarios and of reductions in the baseline level of morbidity, mortality, or vulnerability
- Development and application of systematic standardized methods for assessing cobenefits, including associated economic evaluation
- Development of robust and sophisticated assessment methods for evaluating the health cobenefits and/or adverse impacts of mitigation measures such as biofuels, multimodal transportation, and geoengineering
- Identification of the available resources, limitations of, and potential actions by the current U.S. health care system to prevent,

prepare for, and respond to climate-related health hazard and to build adaptive capacity among vulnerable segments of the U.S. population

Risk assessments are needed to address these aims, including well-established methods—such as time-series studies to describe the current relationships between meteorological variables and health risks—and rapidly developing fields, such as empirical and biological modeling of climatic and other factors affecting the distribution of infectious diseases. Of particular difficulty and importance are hybrid models or protocols that effectively bring these two types of assessments into a common framework.

IMPACTS ON THE ECONOMY OF THE UNITED STATES

The Kyoto Protocol set binding targets for 37 industrialized countries and the European Community to reduce greenhouse gas emissions. President Bush did not support signing the agreement in 2001 because it "would cause serious harm to the U.S. economy."[12] The United States has a new president, and economic impacts of climate change are of high near-term policy relevance. The economic impacts from greenhouse gas mitigation policies are among the most important unknowns in the climate policy debate.

Broadly speaking, there are two mechanisms by which climate change has a fundamental impact on the economy of the United States (and on the world economy). First, economic activities that depend on climate (e.g., agriculture) are affected by a change in the climate, and for large climate changes, that effect will undoubtedly be negative. Whatever the damages, they are expected to rise rapidly as the magnitude of climate change increases. For example, one study (Nordhaus, 2008) estimates that the annual economic damages from a 2.5°C temperature increase are only 20 percent of the annual damages from a 6°C temperature increase. There is, of course, a great deal of uncertainty in the likely damage caused by climate change, as discussed below. These damage estimates typi-

---

[12] Letter from G.W. Bush to Senators Hagel, Helms, Craig, and Roberts, March 13, 2001, available at *www.whitehouse.gov/news/releases/2001/03/20010314.html.*

cally assume that adaptation will be pursued to soften the potential impacts of climate change. Adaptation itself will involve costly actions, but if adaptation is not pursued, damages would be higher. For larger increases in temperature, the bulk of the damage is expected to occur through unanticipated and abrupt change.

One of the biggest impacts from climate change on the economy of the United States will be through coastal flooding. The damage and disruption that accompanies hurricanes and other severe weather events will be magnified by a rise in sea level. Nearly every sector of the economy as well as the welfare of individuals will be affected in some way by climate change.

The second way that climate change will affect the U.S. economy is the cost of reducing greenhouse gas emissions—mitigation. Although there may be pleasant surprises as emissions of greenhouse gases are reduced (such as energy-saving innovations or companies that do better than expected in achieving reductions), there will be costs and those costs will be borne by everyone. Higher prices for energy and energy-intensive goods are usually needed to reduce consumption. People will reduce their energy consumption and carbon generation, but not entirely painlessly. The more slowly emissions are reduced, the easier it will likely be. For instance, allowing more time to reduce emissions avoids premature retirement of energy-inefficient capital. Of course, the down side is the delay in reducing emissions. A primary tool for evaluating policy options is integrated assessment models (Box 2.10).

Research Needs

Despite work cited here, in IPCC reports, and elsewhere, our knowledge of the economics of climate change is surprisingly incomplete and imprecise. Given that we are making decisions on trillion-dollar investments to control greenhouse gases based on what we know now, it seems clear that gaining a better understanding of the economics of climate change should be a high social priority. Many of the economic research problems associated with climate change can be categorized into five broad issues: mitigation of greenhouse gases, regulatory response, impacts of climate change, incidence, and adaptation. Other issues

---

**BOX 2.10 Integrated Assessment Models of
the Climate and the Economy**

Integrated assessment models have become one of the most useful and well-developed approaches to examining the climate problem and what to do about it. The concept is simple. A pure climate model represents how the climate will evolve given exogenous drivers from the economy, where emissions originate. A pure economic model of climate treats the consequences of emissions as exogenous. An integrated assessment model captures in a compact fashion how the climate evolves in response to emissions, how the changed climate impacts economic activity in the world, and how those impacts in turn are combined with mitigation costs to affect policy and the evolution of the economy. Integrated assessment models differ in the level of detail on climate and/or the economy and in the level of closed feedback between climate evolution and economic evolution.

One of the earliest integrated assessment models was developed in the 1970s by Edmonds and Reilly (1983). Other early examples include models developed by Manne and Richels (1991) and by Nordhaus (1977, 1991). The mid 1990s saw major progress on this front, with the development of the Dynamic Integrated Climate Economy (DICE) model (described in Nordhaus, 1994) and other more advanced models (see review in IPCC, 1995). In the DICE model, the atmosphere is represented by a two-box dynamic model and the economy is represented by a single sector. The decision variables are capital investment and investment in mitigation, and all else flows from them.

Over the past 15 years, a good deal of progress has been made in developing more sophisticated integrated assessment models. The primary advance has been to better represent regional differences in the models, rather than view the world as a single economy with average climate impacts. Furthermore, the number of integrated assessment models has multiplied. A recent comparison of global climate-economy models involved 19 different models and modeling groups (Weyant et al., 2006). Although there are many dimensions on which integrated assessment models can be improved, one of the most important is improved data and understanding related to underlying costs, benefits, and economic processes.

---

that touch on climate, such as discounting and uncertainty and risk, are not discussed here. A review of some of the issues in the economics of climate change can be found in Kolstad and Toman (2005) and Heal (2009).

Mitigation

The cost to reduce carbon emissions in particular sectors by particular amounts is subject to a great deal of uncertainty, in both the short run and the long run, and for consumers as well as businesses. Although there have been a number of studies of this problem, a great deal of uncertainty remains. Some analysts suggest a low or negative cost for significant reductions; others suggest significant positive costs (see the discussion in Fischer and Morgenstern, 2006). For example, the state of California estimated that costs of reducing emissions would be negative (there would be a savings due to mitigation), a result roundly criticized by peer reviewers.[13] A recent comparison of economic models found that the cost of controlling an extra ton of carbon in 2025, assuming policies to limit greenhouse gas concentrations to double preindustrial levels, ranged from $2.8 per ton to $482 per ton (Weyant et al., 2006).

Estimates of the costs to reduce carbon emissions have been produced for the economy as a whole (e.g., Nordhaus, 2008) as well as at the sectoral level, particularly by the IPCC. For example, IPCC (2007d) suggests that substantial emission reductions can be obtained in the building sector at negative cost; other negative cost opportunities exist in other sectors. However, the IPCC estimates are neither specific to the U.S. context nor comprehensive, and they do not deal with the rate of change of mitigation as it affects costs. *The Economics of Climate Change* attempted to quantify both the costs and benefits of mitigation (Stern, 2006). However, the data underlying the analysis are sparse (see Symposium on Stern Review in the Winter 2008 issue of the *Review of Environmental Economics and Policy*).

The United States has successfully reduced emissions of air pollutants such as SO_2 (Ellerman et al., 2000). However, the policy challenge of reducing CO_2 emissions is economically different in two ways: (1) CO_2 emissions come from many diverse sources throughout the economy, whereas the bulk of SO_2 emissions came from a few hundred electric power plants, and (2) behavioral change and technological innovation are both likely to play a more profound role with CO_2 reduction because carbon is integral to fossil fuel,

---

[13] See *http://www.arb.ca.gov/cc/scopingplan/economics-sp/peer-review/peer-review.htm.*

whereas sulfur is a contaminant that can be removed. Therefore, it is important to develop a better understanding of the determinants of behavioral change and technological innovation. A research program including cost engineering studies, econometrics, and field experiments (e.g., artificially changing rate structures and observing how behavior changes) would seek to answer questions such as the short-run and long-run marginal cost of reducing CO_2 emissions by various levels for the automobile industry, and ways to accomplish the reduction most effectively (e.g., by changing vehicle design or the CO_2 content of fuels, reducing miles traveled per vehicle).

Regulatory Response

Emission reduction responses depend on policies such as fuel efficiency standards, fuel taxes, feebates,[14] technology-push regulations,[15] and cap-and-trade systems. Experience with cap-and-trade systems is limited to the European Trading System for Carbon (see papers in the Winter 2007 issue of the *Review of Environmental Economics and Policy*), the U.S. sulfur trading system (e.g., Ellerman et al., 2000), and a number of localized trading systems. We have learned a great deal about these economic incentive systems, although our experience with economy-wide trading systems is limited. Other economic incentives are in use as well as prescriptive regulation (Freeman and Kolstad, 2006) and regulations that rely on voluntary actions (e.g., Morgenstern and Pizer, 2007). Little experience exists for carbon regulation (Box 2.11), which will be fundamentally different from many previous regulatory regimes in that behavior as well as technology will be affected. For example, if electric utilities face a price of carbon permits equal to $100 per ton of CO_2, what investments in renewable energy can be expected? How will drivers and automobile manufacturers respond to an upstream (regulation at the energy producer) cap-and-trade system versus a carbon tax or a downstream (regulation at the energy consumer) cap-and-trade system with a similar carbon price?

---

[14] A feebate involves a rebate to above-average performers, financed by a fee on below-average performers, so that no net revenue is collected.

[15] A technology-push regulation is one designed to spur innovation and expand the menu of technological options.

BOX 2.11 The Energy Price and Economic Effects of Reducing U.S. Carbon Emissions

A number of bills have been introduced in the U.S. Congress to limit the emissions of greenhouse gases over the coming decades (e.g., see Appendix A). As of January 2009, none have passed, in part because of questions about how much reducing greenhouse gas emissions will cost and what will happen to energy prices as a result. Virtually all the proposed legislation relies in large part on a cap on emissions of greenhouse gases nationwide, implemented through a system of tradable emissions allowances. A leading proposal in the most recent (110th) Congress was the Lieberman-Warner Climate Security Act of 2007. One of its key features was a cap-and-trade system, capping greenhouse gas emissions 7 percent below 2006 levels beginning in 2012, gradually tightening to 29 percent below 2006 levels by 2030. A detailed analysis by the U.S. Energy Information Administration found that most of the emission reductions would come from the electric power sector via changes in the way electricity is generated (EIA, 2008a). The effects on price would be too modest for consumers to strongly reduce energy consumption. Gasoline prices were assumed to be 10 to 20 percent higher in 2020 and 20 to 40 percent higher in 2030 than in the reference case. Although these are not trivial increases, they are within the variation in prices consumers experienced in 2008. Losses in total national economic output (gross domestic product) would be less than 1 percent in 2030.

The EIA (2008a) analysis reports precise dollar figures for the consequences of reducing greenhouse gases, but there is considerable uncertainty regarding many of the assumptions and conclusions emerging from this report and others like it. The critical nature of the potential impacts on the U.S. economy illustrates the importance of research to better understand the economic impact of greenhouse gas regulations.

Damage from Climate Change

Costly damages to society are among the consequences of climate change. These costs are poorly understood from a physical point of view, let alone an economic point of view. Figure 2.8 summarizes several studies of the damage to the overall economy from a change in the global mean temperature. It is important to emphasize that the figure suggests more precision in these estimates than is warranted. For instance, for moderate temperature changes (e.g., less than 3°C), the estimates are similar, suggesting consensus. However, there is little consensus regarding the damage from modest climate change. In fact, the degree of uncertainty of climate impacts on the economy is generally considered to be very large.

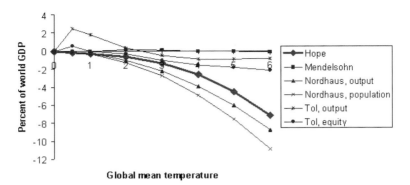

FIGURE 2.8 Some estimates of the global damage from a change in the global mean temperature, as used in several integrated assessment models of climate policy. SOURCE: Dietz and Stern (2008). Reproduced by permission of Oxford University Press. Adapted from Smith et al. (2001), Figure 19-4.

A number of studies have focused on impacts for individual economic sectors. Agriculture costs have been studied most, but many important sectors of the economy have received virtually no attention (Mendelsohn et al., 1994; Mendelsohn and Neumann, 1999). The effects of warming can be mixed, bringing benefits in some cases and costs in others. If the temperature rises when it is cold, there can be less crop damage from freezing, less energy needed for heating, and fewer deaths from cold. However, if the temperature rises too much when it is already hot, crop damage can be severe, more energy is needed for air-conditioning, and some will die from heat waves. The net impact of a given climate change scenario can therefore be quite ambiguous. There is some evidence that substantial damages from climate change may be associated with extreme weather events, but such events (by definition) are rarer and less well studied in both the natural and social sciences (see the "Extreme Weather and Climate Events and Disasters" section). Understanding the damages from temperature extremes is a crucial issue that will likely require more refined spatial and temporal detail than exist in most datasets.

Incidence

Aggregate net benefits (market plus nonmarket environmental benefits minus market plus nonmarket costs) is not the only metric to use in evaluating greenhouse gas policies. The distribution of net benefits and costs from controlling greenhouse gas emissions or from the impacts of climate change and adaptation also has ramifications for environmental justice. For instance, with a cap-and-trade system covering the entire U.S. economy, what income groups end up paying for the costs of greenhouse gas regulation and where do job losses and gains occur? Although some work has been done on who ultimately pays and/or benefits (the incidence) from environmental regulations generally (Metcalf, 1999; West and Williams, 2004), this topic remains largely uninvestigated. This literature generally finds carbon taxes to be moderately regressive.

Greenhouse gas regulation will reduce energy consumption and thus, in all likelihood, emissions of associated non-greenhouse-gas pollutants. The levels of changed emissions of these copollutants are poorly understood as are the monetary benefits of the decreased levels of copollutants (the cobenefits). For instance, what reduction in conventional air pollutants can be expected in urban areas as a result of greenhouse gas regulations? Although some work has been done on this question (e.g., Wier et al., 2005), research is needed to better understand the interplay between co-pollutants and greenhouse gases from a regulatory perspective.

Adaptation

Economic analyses outside the climate arena consider adaptation primarily in the context of price changes. When the price of gasoline goes up by $1, people may adapt by driving a more fuel-efficient car, moving closer to work, or modifying their driving habits. One of the earliest papers on the economics of adaptation focused on investments in irrigation as a way of adapting to uncertainty over precipitation (McFadden, 1984). Such defensive expenditures can blunt the damage from climate change. The nature of the adaptation depends on the speed of the change.

People and businesses will similarly adapt to climate change (e.g., Reilly and Schimmelpfennig, 2000; Kelly et al., 2005; IPCC, 2007c; Mansur et al., 2008), although the magnitude and speed of that adaptation are not well understood. When farmers perceive a changed climate, they will change their agricultural practices; when individuals see their local climate become less hospitable, they may migrate to better climes. This aspect of adaptation is autonomous, since it will occur naturally without government intervention. In contrast, changing power lines, water systems, levees, and other public infrastructure to withstand climate change may require complex government action and thus governmental planning and decision making. This sort of adaptation can be called public adaptation and it will not occur automatically. Research is needed to better understand both the private and public adaptation processes so we can better estimate the costs and damage from climate change policies. Furthermore, the timing of public adaptation is important for public policy.

WHERE DO WE GO FROM HERE?

Climate change is having an impact on basic human requirements, such as water, food, and health. These impacts will become larger in the coming decades. A research program that integrates across the many dimensions of this issue is needed (1) to guide the nation in the multiple choices it faces to reduce the costs and risks of these impacts, and (2) to provide early warning of changes that are abrupt and large enough to push climate and human systems past tipping points. The nation must prepare itself for the possibility of warming in excess of 3°C by the end of the century, followed by the disappearance of most alpine glaciers, the rapid disintegration of the Greenland Ice Sheet, and a rise of sea level of up to several meters (cf., Figure 2.1). It must also prepare for intense severe weather and heat waves, which stress the nation's ability to provide needed water supplies. Such stresses need to be considered in the context of other stresses almost certain to be occurring, such as economic changes, changes in the global market, and potential international conflicts. Preparations will require the integration of models and observations at a much more advanced level than is possible now, as well as the knowledge that comes from linking research on the natural climate system with research on human drivers and responses, and factoring in the needs of decision makers in the research agenda. This in turn requires maintaining a strong natural science research component while strengthening human dimensions research and developing more fruitful interactions with decision makers. The societal issues discussed above provide a framework for human dimensions research. But given the historic emphasis of the program on the natural sciences, a focused effort on key aspects of the human dimensions is also needed to speed progress and further develop the research priorities. The key elements of a research program aimed at understanding climate change and supporting climate-related decisions are discussed in Chapter 3.

3

Future Priorities

The committee was charged to identify priorities to guide the future evolution of the Climate Change Science Program (CCSP), including changes of emphasis and identification of program elements not supported in the past. A long list of priorities was identified from reports, assessments, and workshop discussions, as described and summarized in Appendix C, and from the common themes and gaps that emerged from the research needs outlined in Chapter 2. From these, the committee identified a small set of priorities for the program as a whole. This chapter discusses these priorities in the context of the major roles of a federal climate change research program (Box 3.1). No attempt was made to lay out a comprehensive agenda in any of these areas. Rather, the focus is on what adjustments should be made to the future program to facilitate the integrated, end-to-end approach described in Chapter 2. A key assumption was that energy and geoengineering and other technologies for mitigating climate change research would continue to be primarily the mandate of partner programs such as the Climate Change Technology Program (CCTP).

BOX 3.1 Roles of a Federal Climate Change Research Program

The roles of a federal climate change research program are to

1.   Coordinate federally-sponsored research on climate, human, and related environmental systems across multiple agencies to strengthen synergies and find efficiencies
2.   Develop a research program and a strategic planning process to identify critical gaps and emerging issues and to secure the necessary resources to address them
3.   Ensure the availability of climate-quality observations and computing capacity and the development of human resources and institutions needed to address key priorities
4.   Support coordinated U.S. participation in international climate science initiatives, including global observation networks and international assessments
5.   Facilitate and, where appropriate, leverage regional, state, and local research on climate change, including monitoring and understanding the effects of adaptation and mitigation
6.   Communicate reliable, unbiased research findings and information needed to improve public understanding of climate change and support informed decisions on adaptation and mitigation

---

Much has been written about programs that are needed to implement the various roles listed in Box 3.1. Principles and recommendations on improving management and strategic planning (role 1) for the CCSP are discussed in NRC (2004c) and NRC (2005b). Below we discuss the management challenges that a coordinated multiagency program will face as it moves toward building the knowledge needed to inform decisions. The biggest research gap in the current program (role 2) concerns the human dimensions of global change (e.g., NRC, 1992, 2004c, 2007c), and the discussion below focuses on the importance of adaptation, mitigation, and vulnerability research to support the scientific-societal issues outlined in Chapter 2.

Priorities for space-based observations (part of role 3) for the National Aeronautics and Space Administration (NASA) and the National Oceanic and Atmospheric Administration (NOAA) are identified in the National Research Council's (NRC's) Decadal Survey (NRC, 2007b). This chapter discusses observations that were not included in the Decadal Survey but are needed to understand the climate–human–environment system, as well as data collection issues relevant to a multiagency program. The chapter also discusses other aspects of roles 3 (e.g., modeling and computation), 4 (international partnerships), and 5 (state and local government partnerships) needed to promote an end-to-end approach to climate change. Finally, communications and decision support (role 6) are discussed in NRC (2007c) and NRC (2009), respectively. Below, we focus on only one aspect of this latter issue—climate services— which is under active discussion in Congress and by the CCSP agencies.

CLIMATE OBSERVATIONS AND DATA

Observations are the foundation of climate change research programs. Climate observations and associated climate data records are used to improve our understanding of processes, to monitor the changing climate, to understand how the natural and social systems interact and how these interactions contribute and respond to climate change, and to evaluate the effectiveness of policies to mitigate, cope with, and adapt to climate change (e.g., NRC, 1999a, 2000). The observational components needed for climate research and applications, including ground-based and satellite measurements and socioeconomic surveys, are collectively referred to as a climate observing system.

NASA's Earth Observing System (EOS), designed in 1988, is the closest thing the United States has had to the satellite component of a climate observing system. Originally conceived as three series of satellites to provide sustained, long-term measurements of physical climate and other global variables—complemented by ground-, aircraft-, balloon-, and ship-based measurements (ESSC, 1988)—the project was greatly scaled back. In the end, only the first series of satellites were flown and several planned variables (e.g., those related to geological processes) were never measured. Nevertheless, the data from the EOS satellites, as well as myriad remote-sensing and in situ observing programs operated by other agencies and countries, provided the foundation on which many CCSP successes were built (NRC, 2007c, 2008a).

The need for a systematic and comprehensive approach to collecting climate observations has taken on new urgency with the cancellation, delay, or degradation of existing and planned satellite and in situ observing systems and the decreasing budget for observations experienced over the past several years (e.g., NRC, 2007b, c). As stated in this committee's first report, "the loss of existing and planned satellite sensors is perhaps the biggest threat to the Climate Change Science Program" (NRC, 2007c). A coordinated effort to collect long-term, climate-quality data on land and in the oceans and atmosphere is needed to support climate change science. In addition, the need to address climate change issues in the context of mitigation and adaptation has increased the importance of collecting socioeconomic and health data that can be used to understand human drivers and responses to climate change.

Recommendation. At the earliest opportunity, the restructured climate change research program should set the requirements for a U.S.-operated climate observing system and work with participating agencies (federal, state, local, and international) to establish and maintain the system.

Responsibility for observations is distributed across different federal agencies that participate in the CCSP. The program thus is a logical vehicle for developing a climate observing system. The participating agencies will have to design the system and determine their roles and responsibilities for making the observations and archiving and distributing data (NRC, 1999a). The program would have to (1) identify and prioritize the physical, biological, and social science observations needed to support climate change research and applications;[1] (2) advocate for necessary funding; and (3) coordinate with complementary efforts of U.S. state government agencies (e.g., state mesonets participating in the National Integrated Drought Information System) and international programs (e.g., Global Climate Observing System [GCOS], Global Earth Observing System of Systems [GEOSS]) to leverage investments and work toward a comprehensive international global climate observing system (e.g., as called for in NOAA, 2001; GCOS, 2003, 2004; CEOS, 2006). An enormous amount of work

---

[1] A CCSP interagency working group has begun this process, but had not completed it at the time of writing.

exists to draw on (e.g., see references throughout this section). For example, the GCOS program has developed a set of observation requirements and essential climate variables (GCOS, 2006). More recently, priority satellite observations have been identified for NASA and NOAA, as discussed in the next section. The priority missions for 2013 and beyond will need to be reassessed once a comprehensive set of satellite observation requirements have been identified by the restructured climate change research program.

Decadal Survey

The NRC Decadal Survey identified high-priority space missions to support research and monitoring of the Earth system from 2010 to 2020 (Table 3.1; NRC, 2007b). The chapter on climate variability and change identified a set of climate-mission priorities and pointed out the shortcomings of the National Polar-orbiting Operational Environmental Satellite System (NPOESS), which is intended to form the basis for climate observations in the post-EOS era. It found that NPOESS would lack the capabilities of the EOS satellites and that delays and the cancellation of several key sensors would further weaken observing capabilities and introduce substantial gaps in key variables (NRC, 2007b). A subsequent NRC report evaluated which of the original NPOESS sensors were most important to be preserved and gave highest priority to continuity of microwave radiometry, radar altimetry, and Earth radiation budget measurements (NRC, 2008b). Neither report addressed the need for systematic moderate-resolution land surface observations beyond the Landsat Data Continuity Mission as a priority (see "Agriculture and Food Security" in Chapter 2 for a discussion of the need for improved temporal, global coverage at that resolution).

The Decadal Survey focused on the physical Earth system, although some of the proposed missions identified in chapters on land-use change, earth science applications, human health, and water resources may have relevance to mitigation and adaptation. A decadal survey process focused on societal issues could be a useful way for the restructured climate change research program to identify climate observation priorities for (1) in situ land and ocean measurement systems and (2) data on the human dimensions of climate change.

TABLE 3.1   Satellite Measurements Recommended in the Decadal Survey

Agency	Mission Description	Cost ($M)[a]
2010–2013		
NASA	Solar and Earth radiation; spectrally resolved forcing and response of the climate system	200
	Soil moisture and freeze-thaw for weather and water-cycle processes	300
	Ice sheet height changes for climate change diagnosis	300
	Surface and ice sheet deformation for understanding natural hazards and climate; vegetation structure for ecosystem health	700
NOAA	Solar and Earth radiation characteristics for understanding climate forcing	65
	High-accuracy, all-weather temperature, water vapor, and electron density profiles for weather, climate, and space weather	150
2013–2016		
NASA	Land surface composition for agriculture and mineral characterization; vegetation types for ecosystem health	300
	Day/night, all-latitude, all-season CO_2 column integrals for climate emissions	400
	Ocean, lake, and river water levels for ocean and inland water dynamics	450
	Atmospheric gas columns for air quality forecasts; ocean color for coastal ecosystem health and climate emissions	550
	Aerosol and cloud profiles for climate and water cycle; ocean color for open ocean biogeochemistry	800
NOAA	Sea-surface wind vectors for weather and ocean ecosystems	350

TABLE 3.1 Continued

Agency	Mission Description	Cost ($M)[a]
2016–2020		
NASA	Land surface topography for landslide hazards and water runoff	300
	High-frequency, all-weather temperature and humidity soundings for weather forecasting and sea-surface temperature	450
	High-temporal-resolution gravity fields for tracking large-scale water movement	450
	Snow accumulation for freshwater availability	500
	Ozone and related gases for intercontinental air quality and stratospheric ozone layer prediction	600
	Tropospheric winds for weather forecasting and pollution transport	650

[a] Rough cost estimates, in FY 2006 dollars.
SOURCE: NRC (2007b)

Human Dimensions Observations

The shortage of reliable and consistent data on the interactions between climate, humans, and environmental systems limits our ability to understand how humans affect climate and vice versa, and hence to design policy responses to climate change. This shortage is particularly critical in less developed regions of the world, where socioeconomic and health data may be absent, unavailable, and/or unreliable. Even in developed countries such as the United States, demographic (e.g., housing, census), transportation, economic, and other observations on humans, organizations, institutions, cultures, and societies are sparse and the associated location information may be unavailable to protect individual privacy. There is a particular need for:

• Time-series data related to human pressures on the environment, such as land cover and land use, resource extraction, energy consumption, pollutant emissions from different sources and sectors, and human attitudes, valuations, and responses

- Data on human exposure, sensitivities, and responses to global environmental change, such as morbidity and mortality associated with air and water quality, and vulnerabilities to extreme weather and climate events

Moreover, human-social variables tend to be measured and the data organized for purposes other than climate change research. For example, the Department of Energy's (DOE's) data on energy consumers in households and businesses are not organized in a way that could support research on the causes and trends of greenhouse gas emissions in the United States (Appendix D). To be most useful for climate research, human dimensions data must be better organized and available at different scales of aggregation, including data from surveys and case-study libraries. Finally, data on human systems are rarely coordinated with other observational systems, making it difficult to carry out global analyses or integrated social-natural systems research. Some of the data to support integrated assessments of climate change and other studies of social and ecological systems are coming from research initiatives such as the National Science Foundation's (NSF's) Biocomplexity Program and its successor Dynamics of Coupled Natural and Human Systems program (e.g., Box 3.2). Such programs show what might be possible for a restructured climate change research program. Major research directions for the human dimensions, which would provide a focus for collecting and organizing observations, are discussed in the section "Human Dimensions of Climate and Global Change Research," below.

ANALYSIS OF EARTH SYSTEM DATA

The climate record is built from the analysis of many types of weather and climate-related observations. High-quality, long-term datasets are critical for making better predictions and hence for developing management scenarios to inform decision making and respond to climate change. However, the shortness and/or inhomogeneity of many climate datasets can limit their usefulness for studying climate variability and change and supporting decision making. The value of diverse atmospheric observations can be

---

BOX 3.2 Carbon Storage in Residential Neighborhoods

Research on human-ecosystem interactions is yielding new insights on how homeowner preferences affect land use and hence carbon storage in exurban (beyond the suburbs) areas.[a] In one project, coupled human-ecological models were built that integrated social data (surveys of over 4,000 residents in southeastern Michigan) with land-use change spatial data (parcel records from municipalities and aerial photographs) and satellite data (Landsat and Advanced Very High Resolution Radiometer). The models showed that exurban development increases vegetation productivity (Zhao et al., 2007) and that residential preferences for landscapes that look like those of their neighbors affect ecological function (Zellner et al., 2008). A follow-up study will examine how zoning and other policies might enhance carbon storage in exurban residential areas. For example, policies advocating increased carbon storage are likely to encourage more vegetation, whereas policies advocating water conservation are likely to encourage less. Because exurban development in the United States and other developed countries covers large areas, local policies and homeowner preferences may have regional- and global-scale implications.

[a] Project SLUCE: Spatial Land-Use Change and Ecological Effects at the Rural-Urban Interface: Agent Based Modeling and Evaluation of Alternative Policies and Interventions. See *http://www.cscs.umich.edu/sluce/*.

---

improved by assimilating them into a global atmospheric model to produce a best estimate of the state of the atmosphere at a given point in time. Such global analyses of atmospheric fields have supported many needs of the research and climate modeling communities. Since they are primarily produced by operational forecasting centers, which are less concerned with long-term data consistency, many changes are made to both the models and the assimilation systems over time. These changes produce spurious "climate changes" in the analysis fields, which obscure the signals of true short-term climate changes or interannual climate variability.

A solution is to reanalyze the diverse atmospheric observations over time using a constant (or "frozen") state-of-the-art assimilation model (e.g., Kalnay et al., 1996; Uppala et al., 2005). Today, the products of these global reanalyses provide the foundation for assessments of the state of current climate; diagnostic studies of weather systems, monsoons, El Niño/Southern Oscillation (ENSO), and other natural climate variations; and studies of climate predictability (e.g., Trenberth et al., 2008). They also support regional reanalysis projects and downscaling for studies of local climate and climate impacts. Moreover, the reanalysis process reveals deficiencies in assimilation and prediction systems that need to be improved. For the detection and attribution of long-term climate trends and variability, the quality of the observations and the data assimilation systems and changes in the number and types of atmospheric observations over time can limit the utility of the atmospheric reanalysis products.

Reanalysis is being extended to support research on other aspects of the climate system. As assimilation techniques for observations of atmospheric trace constituents (e.g., aerosols, ozone, carbon dioxide) are refined, reanalysis should eventually provide the means to develop consistent climatologies for the chemical components of the atmosphere, including the carbon cycle, and thus help to quantify key uncertainties in the radiative forcing of climate (IPCC, 2007a). Reanalysis (or synthesis) of ocean data has led to novel techniques to increase the homogeneity of small historical ocean datasets. Other promising developments are occurring in sea ice and land surface reanalysis, and coupled data assimilation systems are beginning to be developed. Finally, adaptation and mitigation planning requires decadal forecasts of the natural climate variability and the response of the system to future changes in greenhouse-gas, aerosol and land-surface forcing. Coupled analysis and reanalysis products are necessary to provide the initial conditions for developing these decadal prediction systems.

Improvements in reanalysis depend on continued support for the underpinning research, the development of comprehensive Earth system models to expand the scope of reanalysis, and the infrastructure for data handling and processing. As the scope of global reanalysis grows, so will the research effort and the need for international cooperation.

Recommendation.   A restructured climate change research program should sustain production of atmosphere and ocean reanalyses, further develop and support research on coupled data assimilation techniques (e.g., for the land surface), and improve coordination with similar efforts in other countries.

EARTH SYSTEM MODELING

From Global Projections to Regional Predictions

Despite impressive gains in knowledge of global climate change, our predictive capability of the Earth system remains insufficient for many societal needs, particularly for forming adaptation and mitigation strategies, which would benefit from more accurate and reliable predictions of regional climate change (NRC, 2007c). Improved predictions of climate change at regional and local scales should help a restructured climate change research program to bridge the gap between science and decision making.

Improving attribution and regional prediction of weather and climate will require improved numerical models. In particular, a stepwise jump in accurately representing the continuum of temporal and spatial variability arising from a wide range of physical and dynamical phenomena and their associated feedbacks is a challenging but essential goal. Our limited understanding and capability to simulate the complex, multiscale interactions intrinsic to atmospheric, oceanic, and cryospheric fluid motions is a barrier to advancing weather and climate prediction on timescales from days to years.

The leading-edge need is to develop a more unified modeling framework that provides for the hierarchical treatment of climate and forecast phenomena that span a wide range of space and timescales. To plan for the effects of climate change, the next generation of global climate models will have to provide numerical simulations on a spatial scale of a few kilometers, with enhanced vertical resolution and better representation of the upper atmosphere. For example, the poor representation of cloud processes is currently a major contributor to uncertainty in the response of the climate system to changes in radiative forcing. Such models are essential to improve our understanding of the multiscale interactions in the coupled system, to identify those of greatest importance, and to document their effects on climate. Ultimately, such basic research will help determine how to better represent small-scale processes in climate models; for instance the manner in which moist convection and its associated mesoscale organization drives larger circulations or the complex regional climate processes that occur along the west coasts of continents in tropical and subtropical zones. Another example is the simulation and prediction of hurricanes and depiction of their effects on climate in models, which has been missing altogether.

Sustained, long-term, global observations are needed to develop, initialize, and constrain the models. The distinction between shorter-term predictions and longer-term climate projections is becoming blurred, given the realization that all climate system predictions may require that coupled general circulation models be initialized with best estimates of the current observed state of the atmosphere, oceans, cryosphere, and land surface. However, there are many challenges. For instance, there is currently no direct way to measure soil moisture, and ocean salinity reconstructions remain a significant problem for initializing the ocean circulation. Providing more credible predictions of regional variability and change will therefore require more work on data assimilation techniques and stronger links to numerical weather prediction. In addition, there is a great need to better characterize and quantify the uncertainties in climate system predictions to best guide mitigation policy and adaptation strategies. All of these advances will require more people and more powerful computers dedicated to reliably predict climate and associated uncertainties with a level of detail and complexity that is not possible now.

Recommendation. The restructured climate change research program should develop and implement a strategy to improve modeling of regional climate change. Improved predictions of climate change at regional and local scales will require (1) a new suite of high-resolution climate models; (2) increased computational resources; (3) tighter connections between climate model development, numerical weather prediction, and data assimilation research; and (4) a larger cadre of scientists capable of developing models and analyzing model output at the regional scale.

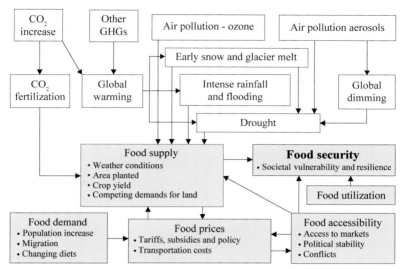

FIGURE 3.1 Multiple stressors related to climate, agriculture, and food security. Biophysical factors (white boxes) and socioeconomic factors (blue boxes) interact in the context of climate change to affect food security. Although food security depends on supply, accessibility, and utilization, food supply is a function of a complex interaction of climate and socioeconomic conditions and trade.

Integrated Modeling of Multiple Stressors

Climate change is occurring in concert with other environmental and socioeconomic changes. A better understanding of the interactions and feedbacks between different components of the natural and social systems is required to understand the potential impacts of climate change and the various responses. Models are a primary means for understanding such processes and assessing potential outcomes and policy options. However, few models are capable of simulating the complex interplay of multiple stressors (e.g., energy use, land-use change, water resource availability, societal risks and vulnerabilities; see Figure 3.1) on environmental and social systems. Research challenges include the integration of models and datasets with inherently different spatial and temporal scales and uncertainties, and the limitations of socioeconomic datasets, which are often available only in aggregated form (see "Climate Observations and Data," above).

Interest is growing in coupling assessment models with Earth system models. For example, the CCSP used three integrated assessment models to calculate mitigation costs of emission scenarios (Levy et al., 2008). A recent workshop addressed the changing role of integrated assessment models in the context of climate change mitigation and adaptation (DOE, 2008). The workshop identified areas of emphasis for impacts modeling (e.g., water quantity, quality, supply, and management) and for exploring the relationships between climate and the energy and transportation sectors. It also laid out a research agenda for the adaptation domain, starting with addressing decisions in the agricultural, energy, and forestry sectors, for which the integrated assessment models already have reasonably sophisticated representations.

Only integrated models can reasonably be expected to explore the quantitative relationships between decision making on mitigation and adaptation, and influences on the carbon cycle and other aspects of the environment. A much larger investment in integrated model development, validation, uncertainty analysis, and model intercomparison is needed to achieve the potential payoffs.

Recommendation. The restructured climate change research program should support advances in integrated modeling to address science and policy questions associated with the impacts of climate change and mitigation and adaptation responses. In particular, such integrated models should be used to improve the characterization of the end-to-end uncertainty in projected climate changes and impacts at regional and local levels.

HUMAN DIMENSIONS OF CLIMATE AND GLOBAL CHANGE RESEARCH

The biggest shift in emphasis for the restructured climate change research program is to give considerably more attention to the human dimensions of climate change, a research element of both the CCSP and its predecessor U.S. Global Change Research

Program (USGCRP) that has been significantly underfunded in the past (NRC, 2004c, 2007c). Human dimensions research seeks to answer questions about the role of human actions and behavior in changing climate and in mitigating and adapting to the impacts of climate change. Despite the importance of these issues, however, spending on human dimensions research has never exceeded 3 percent of the CCSP research budget (NRC, 1992, 2007c). As a result, research, data collection, and modeling of socioeconomic and behavioral functions have lagged behind corresponding activities on the physical climate system, and the human capacity has fallen short of what is now needed (NRC, 2007c). Without adequate research and capacity, the provision of usable science for policy and decision making will be severely limited.

Chapter 2 lays out a research agenda in which human dimensions and natural science are integrated to address societal issues. Progress on these issues would be sped by also making a focused effort to strengthen research in the following areas: (1) understanding and quantifying societal gains and losses from taking or not taking action, (2) human drivers of climate change, (3) vulnerability and adaptation, and (4) mitigation. The last three are described below and in greater detail in Appendix D, and the first, which came out strongly at the committee's second workshop, is described in "Impacts on the Economy of the United States" in Chapter 2. Knowledge of human driving forces, vulnerability, impacts, and responses is also needed to improve the integrated assessment models discussed above. Until these variables can be represented realistically, the models will be insufficient and incomplete.

Research on the human drivers of climate change seeks to understand how humans affect rates of greenhouse gas emissions through population growth, migration, behavior, technological change, land use, or consumption (e.g., NRC, 1997, 1999c, 2005a; Kates, 2000). It examines how behavior at the individual, household, and organization levels drives climate change and how institutions and governance both shape these drivers (e.g., by affecting resource use) and create possibilities for mitigation and adaptation. Specific topics to support policy include the factors that influence population and consumption growth; the links among economic consumption, resource consumption, and human well-being, including the potential to satisfy basic needs and other demands with significantly less resource consumption; and ways that population growth and consumption behavior responds to efforts to change them through information, persuasion, incentives, and regulations (Appendix D). These behaviors occur in the context of the natural environment, and so research is also needed on the interactions between natural and social systems and the net effects of population growth and human behavior on water, land, and energy use; carbon fluxes; ecosystems; coastal resources; and the built environment.

Studies of vulnerability and adaptation focus on the material conditions, values, institutions, governance, and politics that shape individuals' and organizations' vulnerability (exposure and sensitivity), adaptive capacity, and adaptation options and barriers, and their ability to cope with and recover from the impacts of climate change. Adaptation refers to social and economic changes undertaken in response to climate change impacts. Such changes may be autonomous (triggered by other ecological, market, or welfare changes) or planned (deliberate policy decisions aimed at returning to, maintaining, or achieving a desired state of affairs) (IPCC, 2007c, Appendix I). Adaptation may be anticipatory when it seeks to prevent and prepare for, rather than respond to, an actual change.

There are many constraints on adaptation, particularly on organized adaptation aimed at social change. For example, the political context for adaptation must be considered. Climate is but one of many issues that come before Congress and state legislatures, and its perceived priority is often lower than that of issues such as national security or the economy. Enabling or fostering adaptation by enacting new laws or amending existing statutes requires not only the political will to move forward, but also a time-consuming balancing of interests through the political process. Another barrier is that social networks connect many of the early and later adopters in adaptation structures in which communication and information is often slower and more uneven than rapid adaptation demands. For example, although seasonal climate information is widely and rapidly disseminated, organizational and institutional constraints can inhibit its use (Beller-Simms et al., 2008). Many behavioral constraints (e.g., established roles, professional training, bureaucratic inertia) are not well understood and are entrenched in political and economic structures and practices (Lemos, 2008). A related problem is that climate products developed by scientists in isolation from information users commonly do not meet managers' needs, preventing their use for adaptation.

Vulnerability is the degree to which the environmental or human systems are unable to cope with the adverse effects of climate change and experiences harm. Integrated research to find robust approaches to support policy design and implementation to decrease vulnerability includes (1) developing scenarios, vulnerability maps, and adaptive capacity metrics; (2) modeling feedbacks and nonlinearity between adaptation and mitigation; and (3) examining vulnerability, adaptive capacity, and adaptation options on several dimensions, including type of event (e.g., storm surge, crop failure), location and scale, socioeconomic characteristics of affected populations, sector (e.g., water, health), and constraints and opportunities for governance and policy implementation. Other research needs include the evaluation and costing of impacts, mitigation, and adaptation options and a better understanding of climate impacts. For example, estimates of the time trajectories of vulnerabilities could yield scenarios of vulnerability and adaptive capacity that could be integrated with climate scenarios to improve projections of the impacts of climate change (NRC, 1998, 1999c; Appendix D).

Mitigation refers to purposeful efforts to reduce greenhouse gas emissions or enhance greenhouse gas sinks. Mitigation research seeks to understand how the incentives and regulations to reduce carbon consumption will be implemented, how much implementation will cost, and how institutions shape the incentive environment within which mitigation occurs. A range of opportunities for mitigation exist, from human needs and desires to the consequences of climate change (Hohenemser et al., 1985), although all will not be equally cost-effective. Robust mitigation strategies typically rely on risk research and assessment, as well as learning from experience.

In the climate change arena, where global equity issues and international agreements are involved, national programs are not sufficient. For example, in forest-rich countries such as Brazil, the Democratic Republic of Congo, and Papua New Guinea, potential solutions to deforestation (e.g., market-led conservation) are constrained by the lack of baseline data and empirical research on market mechanisms and local governance, and poor understanding of carbon economy institutions that could shape current patterns of land use and change in these countries. On the response side, research is needed to improve our understanding of the many available options, ways to evaluate them across different dimensions (e.g., dollars, species, lives), ways to diffuse them across society, and ways they interact and feed back on each other. For example, the costs and benefits of adaptation may depend on the outcomes of prevention efforts, and both may be affected by the temporal and spatial scale of the analysis (Appendix D). Better understanding of responses is important not only for adverse impacts that are predicted but also for those that have not yet been identified.

Recommendation. The restructured climate change research program should support new research initiatives on (1) human drivers; (2) impacts, vulnerability, and adaptation; (3) mitigation and responses; and (4) understanding and quantifying societal gains and losses.

Over time, these initiatives would help address societal concerns of direct relevance to the program and provide a concrete focus for collecting human dimensions data and growing the research community.

DECISION SUPPORT

A key provision of the U.S. Global Change Research Act of 1990 is to produce "information readily usable by policymakers attempting to formulate effective strategies for preventing, mitigating, and adapting to the effects of global change." The committee's first report (NRC, 2007c) found that use of CCSP-generated knowledge to support decision making and to manage the risks and opportunities of climate change is proceeding slowly. Congressional legislation under discussion would amend the Global Change Research Act to require more focus on science that supports decision making or authorize new research programs on sector-based mitigation or adaptation (Appendix A). A wide variety of policy makers and other stakeholders are making decisions on climate change mitigation and adaptation, including

- State climate coordination groups focused on carbon sequestration and water issues
- State-level managers concerned with natural resource issues, such as water, agriculture, fire, rangelands, and forestry, and with human health
- Federal land and water managers from agencies such as the U.S. Department of Agriculture (USDA), U.S. Forest Service, National Park Service, Department of the Interior, Bureau of Land Management, Bureau of Reclamation, and U.S. Army Corps of Engineers
- Nongovernmental organizations concerned with conservation, policy, and community advice
- Policy makers, including governors, mayors, and county supervisors
- Federal, state, and county health departments
- Private companies and foundations offering products and services related to energy, reinsurance, finance, engineering, agriculture, fisheries, forestry, range management, health, and tourism
- Individuals making climate-related decisions, such as planting drought-resistant crops and consuming water and energy

These stakeholder groups use and/or provide climate-related information, research, and services, often without interacting with the CCSP. The deficiency of two-way communication between the program and stakeholders is a major obstacle to decision support (NRC, 2007c). Engaging stakeholders in a restructured climate change research program would increase the resource base (people, ideas, dollars) to support actions to mitigate and adapt to climate change, inform the program and its researchers about stakeholder priorities, and possibly provide opportunities for leveraging research funding (e.g., California climate research; see Box 3.3). A logical avenue for developing partnerships is through decision support activities, where policy makers and managers have defined

---

**BOX 3.3 California Actions on Climate Change
Adaptation and Mitigation**

The state of California has taken a leadership role in climate change mitigation and adaptation, establishing policies and taking action well in advance of the federal government in many areas. For example, California is the first state in the nation to have adopted a legislatively required greenhouse gas mitigation plan that involves a wide range of economic sectors and includes actions such as

• Establishing a cap-and-trade program that links with seven western states and four Canadian provinces
• Achieving a statewide renewable energy mix of 33 percent
• Establishing targets for reducing transportation-related greenhouse gas emissions, including setting state vehicle emissions standards that are stricter than federal requirements
• Expanding and strengthening energy efficiency programs (CARB, 2008)

Implementing this plan is expected to have significant economic implications, and there are opportunities for directed research to help support implementation. After the federal government, California is the largest governmental funder of climate change programs and supporting research in the nation. The California Energy Commission's Public Interest Energy Research program, funded at $83.5 million per year, supports climate monitoring, analysis, and modeling; improvement of greenhouse gas inventory methods; options to reduce greenhouse emissions; and impacts and adaptation. Work on the latter has included downscaling results of global climate models and developing sector-specific information on impacts at state or regional scales for state agencies to used in adaptation plans required by a governor's executive order.[a]

SOURCES: Hanemann (2008); *http://www.energy.ca.gov/research/index.html.*
[a] *http://gov.ca.gov/index.php?/executive-order/1861/.*

---

goals, such as compliance with statutory mandates, that could inform the research agenda.

The CCSP has taken the first steps toward supporting decision makers through pilot programs of individual agencies. These programs range from providing information and tools needed by a variety of stakeholder groups (e.g., National Integrated Drought Information System, seasonal outlooks, Environmental Protection Agency's [EPA's] National Center for Environmental Assessment) or specific sectors (e.g., NOAA's Sectoral Applications Research Program), to actively engaging with stakeholders to determine their needs and provide tailored information products and services (e.g., NOAA's Regional Integrated Sciences and Assessments [RISA] program, International Research Institute for Climate and Society).[2] The RISA program in particular has had successes in delivering information stakeholders need far out of proportion to its modest funding (about $6.6 million annually), earning the support of some stakeholder groups.[3] Although these programs have proven useful, they are small and ad hoc (NRC, 2007c).

An NRC report lays out a comprehensive framework for organizing climate-related decision support activities, including principles for effective decision support, provision of climate services, and research needed to support the services (Box 3.4; NRC, 2009). The report recommends that decision support activities be carried out by organizations that are closest to users, including federal, state, and local government agencies and private organizations. Federal roles would include (1) supporting decision making by federal agencies and the constituents they are bound by statute or mandate to serve, and (2) facilitating the development and improvement of decision support systems by nonfederal entities by providing scientific research, methods, communication networks, databases, standards, and training. The ultimate objective would be to create a distributed capacity for decision support that helps decision makers better cope with surprise and local climate change conditions.

Such a distributed capacity for decision making raises challenges for research. We currently possess only limited knowledge of how such decisions may best be made and when decisions may be better deferred in hopes that uncertainties will be narrowed by further research. As pointed out in a number of IPCC reports, current uncertainties about climate change will not be easily resolved by research carried out now or in the near term. Nevertheless, decisions will have to be made. Basic research, such as that sponsored

---

[2] See *http://www.drought.gov/portal/server.pt/community/drought.gov/202,*
http://cfpub.epa.gov/ncea/cfm/recordisplay.cfm?deid=157003,
http://www.climate.noaa.gov/cpo_pa/sarp/,
http://www.climate.noaa.gov/cpo_pa/risa/,
http://portal.iri.columbia.edu/portal/server.pt.
[3] For example, see
http://www.westgov.org/wswc/050407%20risa%20resolution.pdf.

BOX 3.4 Elements of a Decision Support Framework Recommended in
***Informing Decisions in a Changing Climate* (NRC, 2009)**

Principles for effective decision support:

1. Begin with users' needs, identified through two-way communication between knowledge producers and decision makers
2. Give priority to process (e.g., two-way communication with users) over products (e.g., data, maps, projections, tools, models) to ensure that useful products are created
3. Link information producers and users
4. Build connections across disciplines and organizations
5. Seek institutional stability for longevity and effectiveness
6. Design for learning from experience, flexibility, and adaptability

Components of a National Climate Decision Support Initiative:

1. Services, including activities, consultations, and development of decision support networks and processes to identify information needs, provide needed information, and facilitate decision making and learning processes in constituencies affected by climate change
2. Research for informing climate change response, a component of equal importance to current research on climate change processes:
    a. Science to support decision making, including understanding climate change vulnerabilities, mitigation potential, adaptation contexts and capacities, the interaction between mitigation and adaptation, and emerging opportunities associated with climate variation and change (e.g., alternative energy development)
    b. Research on decision support, including research to understand information needs, climate risk and uncertainty, processes related to decision support, design and application of decision support products, and assessment of decision support experiments

---

under NSF's Decision Making Under Uncertainty program, continues to be a pressing need.

A comprehensive decision support framework is described in NRC (2009). The component that is receiving the most attention from Congress and the CCSP is a national climate service, which may be created within NOAA (S 2307) or as an interagency effort. NOAA is currently implementing a recommendation of its Science Advisory Board to examine alternative ways of managing a national climate service.[4] The relationship between a national climate

---

[4] *http://www.sab.noaa.gov/Reports/Reports.html.*

service and a restructured climate change research program is discussed below.

Climate Services

A national climate service could facilitate two-way dialog with stakeholders and translate scientific and technical information into language that is more easily understood by policy makers and the public. It could be responsible for provision of products (e.g., observations, regional forecasts and predictions), tools (e.g., models, Web services), and outreach and extension services needed to support resource managers and policy makers at the national, state, and local levels (NRC, 2001, 2003; Miles et al., 2006).

The potential relationship between the CCSP and a national climate service is illustrated in Figure 3.2. Climate and decision support research, as well as climate models, observations, and assessments provide the underpinning for climate services, and the demand for services in turn will influence the direction of the climate change research program.

Climate services are not currently part of the CCSP, with the exception of pilot efforts (e.g., RISA programs) noted above. Similarly, the user-driven research needed to expand these exploratory initiatives has received little CCSP attention. For example, a logical extension of ENSO forecasting is climate services related to agriculture and water management practices. Further research is needed on the trade-off between forecast skill and information value and the scientific outputs suited to the needs of water resource managers (Beller-Simms et al., 2008; NRC, 2008c). Research is also needed to extend these forecasts to decadal projections, and to provide services in the context of mitigation and adaptation to long-term climate change.

Whether climate services should be included in a restructured climate change research program or only linked to it is under debate. The need for close linkages with the research program that develops the products and tools and also uses some of them to understand trends and improve predictions argues for incorporating climate services into the research program. On the other hand, the operational nature of the activities, the need for supporting data, models, and research from other sources beyond the CCSP (e.g.,

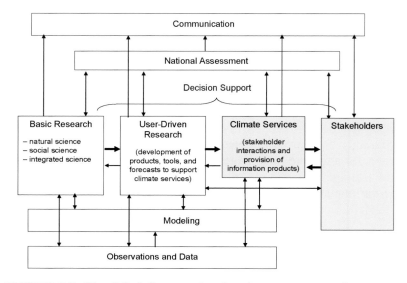

FIGURE 3.2 Simplified diagram showing the components of a restructured climate change research program that is more responsive to the information needs of stakeholders. Basic research underpins the program. Applying the results of basic research to resource management and policy decisions requires user-driven research, and the supporting data collection and model development (e.g., regional models) is influenced by these practical needs. Research results are made available to stakeholders via the communications component of the program. Information products and services targeted to specific user needs (climate services) may be provided outside the program. Climate services, user-driven research, and parts of basic research (e.g., the science of decision support) comprise decision support. Two-way interaction with stakeholders occurs primarily through climate services but also through user-driven research and the national assessment.

CCTP), and the potential to overwhelm the research program with the demands for specialized services argue for establishing a separate climate service. The solution may be to carry out climate service activities outside the program, even if the coordination takes place within the program. A possible model is a cooperative extension service for climate, similar to one established for agriculture by the Smith Lever Act of 1914. The agricultural extension service provides federal support at the state level for meeting public information needs at the local level. However, the best decision support model for any particular case or set of decision makers will be a matter of empirical research (NRC, 2009).

Because successful programs have a leader (NRC, 2005b), the committee recommends that one agency take the lead in developing the climate service, although multiple agencies would have to be involved in its design and implementation. Interagency coordination in the framework of a restructured climate change research program would provide essential linkages to the federal research programs and take advantage of the expertise and capabilities of different agencies and the relationships they have established with the various stakeholder communities. The interagency framework would also provide a mechanism to identify gaps and new priorities and to minimize duplication (NRC, 2009).

Recommendation. The restructured climate change research program provides a framework to coordinate federal efforts to provide climate services to meet the climate information needs of policy and decision makers concerned with impacts, mitigation, and adaptation to climate change at federal, state, and local levels. The services should be led by a single agency but have broad participation from other federal agencies.

NATIONAL ASSESSMENT OF CLIMATE IMPACTS AND ADAPTATION OPTIONS

The 1990 Global Change Research Act calls for a national assessment at least every 4 years

1. To integrate, evaluate, and interpret the findings of the program and discuss the scientific uncertainties associated with such findings
2. To analyze the effects of global change on the natural environment, agriculture, energy production and use, land and water resources, transportation, human health and welfare, human social systems, and biological diversity

3. To analyze current trends in global change, both human-induced and natural, and projects major trends for the subsequent 25 to 100 years[5]

National assessments have the potential to engage stakeholders; to focus research effort on climate change impacts, trends, and predictions needed by decision makers; and to communicate program results. Unlike assessments of published scientific research, such as the Intergovernmental Panel on Climate Change (IPCC) reports, a U.S. national assessment involves undertaking targeted research and creating new datasets and model runs at the regional scale, tailored to address U.S. national issues and concerns. The first national assessment, which was initiated in 1997 and released in 2001, involved 20 public workshops, 5 sector teams, and extensive stakeholder interaction. The resulting report (NAST, 2001) addressed key scientific questions on climate change impacts on the United States. The next U.S. attempt took the form of 21 synthesis and assessment reports on diverse topics identified by the CCSP and an overarching assessment of the effects of global change on the United States. The reports, which were published between 2006 and 2008,[6] were based on findings from CCSP research and previous scientific assessments, particularly the IPCC assessments (CENR, 2008). The overarching assessment was done as a literature review without significant stakeholder involvement. Although the in-depth analysis of specific issues is useful, the collection of disparate reports does not add up to a national assessment, and studying the issues separately misses opportunities to integrate across topics, regions, or sectors (NRC, 2007a).

Ideally, a future national assessment would build ongoing relationships with stakeholders to address evolving scientific and societal needs and to identify useful decision-support products and research priorities. Stakeholder engagement was a strength of the 2001 assessment (Box 3.5). Many of the contacts made in that process will have to be renewed and other individuals with a stake in addressing climate impacts will have to be identified. The sec-

---

[5] P.L. 101-606, 104 Stat. 3096–3104.
[6] The synthesis and assessment reports are available at
http://www.climatescience.gov/Library/default.htm#sap.

toral workshops, convened by CCSP program staff in 2007 and 2008 to seek input on a new strategic plan for the program, were a good step toward building these relationships. The national assessment would act as a catalyst for the development of national and regional data products and models designed to address stakeholder needs. In this context, the assessment will need a strong underpinning of user-driven research (e.g., see Figure 3.2). Emphasis will also have to be given to reporting findings of the assessment, which are often based on complex information, in a way that stakeholders can easily understand. Some form of national climate change indices or report card may provide a useful communication tool. Foci for the next assessment include the following:

- The likely changes over the next few decades, the associated impacts in multiple regions and across various sectors, and mitigation and adaptation options
- The extent to which we understand the science, technologies, economics, and politics underlying mitigation and adaptation strategies in the context of other socioeconomic and environmental changes

Recommendation. The restructured climate change research program should immediately begin planning a national assessment on climate change impacts, adaptation, and mitigation; consulting with stakeholders on the focus, content, and approach; establishing a strategy and schedule for implementation; securing the necessary financial and institutional commitments; and developing the regional climate projections, datasets, and models that will be used.

Depending on the focus, the program may also need to build the scientific capability and human capacity in some areas (e.g., see "Earth System Modeling" and "Decision Support," above).

BOX 3.5 Lessons Learned from the 2001 National Assessment

Strengths of the 2001 National Assessment

- The assessment process was intended to be transparent and in-clusionary
- The assessment engaged a large number of scientists, advanced our understanding of assessment methods, and initiated extensive stake-holder interactions
- Although the questions were mostly framed by policy makers, the results were independent and the conclusions were not subjected to ad-ministrative or policy review

Weaknesses of the 2001 National Assessment

- The process was cumbersome
- Funding for the assessment was not included in the normal budg-eting process, limiting the participation of some agencies
- Private-sector involvement was minimal

Guidelines for a useful assessment

- A clear mandate and well-defined criteria for defining structure and scope
- Strong leadership
- Efficient use of scientific and stakeholder capital (data, people, previous efforts)
- A specific goal of building a community of people and institutions with the knowledge required to work at the interface between basic sci-ence and stakeholders
- A strategy for continued two-way communication between scien-tists and other stakeholders throughout the assessment process
- A commitment to funding

SOURCES: Morgan et al., (2005); NRC (2004c, 2007a); October 2007 workshop.

INTERNATIONAL PARTNERSHIPS

Climate change is a global phenomenon, and a number of countries are investing in climate research, observations, and miti-gation and participating in international climate programs. The participation of U.S. scientists and program managers in setting and helping to implement the research agendas of these international programs strengthens the linkages to the U.S. program and will leverage international investments. Working with the international research community on common problems also increases the pool of scientific expertise and takes advantage of complementary strengths and approaches. For example, the benefits of international cooperation in providing climate services have already been demonstrated by the various Climate Outlook fora, held regularly around the world.[7] International partnerships can also be used to stretch observing system dollars. With the decrease in U.S. funding for Earth observations and the increased investment by nations such as China, Brazil, and India, U.S. scientists will increasingly have to turn to other countries for data. Finally, if the United States is to take an international leadership role on climate change policy, it will need to help the U.S. research community work effectively within international science coordination structures. Strengthening the appropriate program linkages at the international level will help enable the science to inform policy.

The CCSP supports the U.S. contribution to the IPCC, which has played a critical role in developing the international scientific consensus on climate change. U.S. leadership and participation in the IPCC has been substantial. However, the CCSP has not actively coordinated U.S. participation in other international programs that address climate-related research (e.g., WCRP, IGBP, IHDP, IAI, START), assessments (e.g., WHO), or observations (e.g., GEOSS, GCOS, GTOS, GOOS, CEOS),[8] missing opportunities to influence the direction of these programs and find synergies with U.S. programs. Instead, individual agencies have supported the participation of individual scientists in a largely ad hoc fashion. Developing an overall strategy for participating in international programs and sup-

---

[7] *http://www.wmo.ch/pages/prog/wcp/wcasp/clips/outlooks/climate_forecasts.html.*

[8] Note: CEOS = Committee on Earth Observation Satellites; GOOS = Global Ocean Observing System; GTOS = Global Terrestrial Observing System; IAI = Inter-American Institute for Global Change Research; IGBP = International Geosphere-Biosphere Programme; IHDP = International Human Dimensions Programme on Global Environmental Change; START = Global Change System for Analysis, Research, and Training; WCRP = World Climate Research Programme; WHO = World Health Organization.

porting international program offices would help a restructured climate change research program understand the extent of U.S. participation, identify crucial gaps, and set priorities for federal participation in international programs that can help meet its program objectives. A number of international coordination programs are aligning themselves in ways that will facilitate interaction with a restructured U.S. climate change research program. For example, IGBP has added fast-track initiatives to foster integrated research across its core programs (e.g., ocean acidification over time),[9] and GEOSS is organized along many of the themes outlined in Chapter 2 (e.g., health, water, agriculture).

The involvement of the U.S. Agency for International Development in the CCSP has been rather small (about 1 percent of the research budget in 2007; see CCSP, 2008). However, the most vulnerable populations and the largest areas of biodiversity are in developing countries, where climate change will compound other stressors on food and water supply, human health and livelihoods, and biodiversity conservation. As these nations start to respond to climate change impacts and develop adaptation strategies, climate change will have to figure more centrally in the U.S. development agenda (e.g., through participation in the Kyoto Protocol's Adaptation Fund). Nongovernmental organizations with international programs are already developing climate change initiatives in these areas.[10] CCSP research on impacts and adaptation approaches could help guide U.S. investments in developing countries. U.S. Earth observing systems could help target interventions and monitor the effectiveness of these approaches and policies. It is interesting to note that the 2008 drought in Iraq caught the attention of the Department of Defense (DOD), which is concerned with the implications of water scarcity, crop failure, and resulting food shortages on security in the region.[11] Such issues may give DOD a strategic interest in expanding its participation in a climate change research program. The improved regional prediction of floods, droughts, and other extreme events and assessment of their

---

[9] See *http://www.igbp.net/page.php?pid=130.*

[10] See, for example, the World Wildlife Fund's climate program, *http://www.panda.org/about_wwf/what_we_do/climate_change/index.cfm.*

[11] *http://www.mnf-raq.com/index.php?option=com_content&task=view& id=22856&Itemid=128.*

impacts on society may well influence which U.S. agencies are involved in the restructured climate change research program.

Recommendation. The restructured climate change research program should play a lead role in coordinating and increasing U.S. participation in climate-related efforts of international programs, and in developing and implementing a shared agenda of climate observations, research, and applications.

TOP PRIORITIES AND BUDGET IMPLICATIONS

The committee's top priorities, cast as actions for a restructured climate change research program to better meet national needs, are as follows:

• *Reorganize the program around integrated scientific-societal issues to facilitate crosscutting research focused on understanding the interactions among the climate, human, and environmental systems and on supporting societal responses to climate change.* The traditional approach of organizing research along scientific disciplines or themes (e.g., atmospheric composition) cannot fully address issues of concern to society, such as the impacts of severe weather and climate.

• *Establish a U.S. climate observing system, defined as including physical, biological, and social observations, to ensure that data needed to address climate change are collected or continued.* The role of a restructured climate change research program is to develop a prioritized list of satellite and in situ observations and to work with local, state, and federal government agencies and international programs to ensure their collection.

• *Develop the science base and infrastructure to support a new generation of coupled Earth system models to improve attribution and prediction of high-impact regional weather and climate, to initialize seasonal-to-decadal climate forecasting, and to provide predictions of impacts affecting adaptive capacities and vulnerabilities of environmental and human systems.* Achieving this objective requires a considerable expansion of local- and regional-scale modeling activities, supported by advanced computational facilities, and improved and sustained communication with stakeholders.

• *Strengthen research on adaptation, mitigation, and vulnerability.* Integrated research, combining natural, social, and health science from a variety of disciplines, will be required to boost capabilities and enable the results to be applied to a broad spectrum of climate problems. The program will have to find mechanisms for attracting new research talent to build the capacity needed to support sound adaptation and mitigation strategies.

• *Initiate a national assessment process with broad stakeholder participation to determine the risks and costs of climate change impacts on the United States and to evaluate options for responding.* Early planning steps include (1) identifying stakeholders as well as agencies that should be involved so funding can be raised or reprogrammed to ensure their participation, and (2) determining the scope of the assessment so development of the necessary datasets and models can begin.

• *Coordinate federal efforts to provide climate services (scientific information, tools, and forecasts) routinely to decision makers.* Although the design of a national climate service is still under discussion, a restructured climate change research program could begin laying the foundation by identifying the roles of the various federal agencies and increasing emphasis on user-driven research.

All six of these actions are necessary for building a program that supports both research and action. They are listed as a logical sequence of actions, but work can begin on all simultaneously. First is organizing the research around scientific-societal issues to help the program address not only how and why the climate is changing, but also to develop options for adapting to or mitigating the changes. Next is the collection of natural and social science observations to document and understand how the climate is evolving in response to rapid increases in CO_2 and other human drivers. To use these observations to predict future changes, we need new regional- and local- scale models. The observations and new fine-scale models should pave the way for strengthening research on adaptation, mitigation, and societal vulnerability and for undertaking a national assessment on the impacts of climate change.

The data, models, and research results provide the foundation for informing climate change-related decisions, but a new institutional arrangement will be required to work effectively with stakeholders and provide the climate services (specialized products, tools, and forecasts) they need.

Budget Implications

Each of these initiatives would expand the scope of a restructured climate change research program, with varying budget implications. Organizational and planning activities are typically carried out using CCSP Office funding (currently nearly $2 million annually) and the in-kind support of program managers serving on interagency committees. Assuming that such funds continue to be made available, restructuring the research part of the program, setting observations priorities, and planning major initiatives should be cost neutral. Similarly, funding for a national assessment may not require new resources. According to the CCSP Office, the cost of the last national assessment was a few tens of millions of dollars, about the same as the combined cost of the 21 synthesis and assessment reports.[12]

The CCSP budget for FY 2008 was about $1.8 billion (CCSP, 2008). Adjustments on the order of a few tens of millions of dollars should be possible without substantially undermining major parts of the program. Two of the committee's priorities fit into this category. First, increasing research on adaptation, mitigation, and vulnerability would require a substantial increase in funding, but since current funding levels in these areas are low, the total amount would be relatively small. Directing some of the increased funding to support Ph.D. students and postdoctoral fellows in the areas of human dimensions and integrated climate–society systems should encourage growth of this research community. Funding may be available from new partners with expertise in this area (e.g., Bureau of Land Management) or from CCSP agencies that have programs in the human dimensions (e.g., NSF's Social, Behavioral, and Economic Sciences Directorate, DOE's Integrated Assessment

---

[12] Personal communication from Peter Schultz, director of the CCSP Office, on November 20, 2008.

Program, EPA, USDA, National Institute of Environmental Health Sciences; NOAA's RISA program and Sectoral Applications Research Program). Second, more room must be made within the program to expand existing research activities aimed at developing methods and information to support decision making. Programs that are successfully providing prototype climate services (e.g., the NOAA RISAs) are funded at a relatively low level (i.e., less than $10 million per year), and significant increases would not adversely affect the natural science program.

CCSP program activities and associated budgets are reported annually to Congress in *Our Changing Planet* (e.g., CCSP, 2008). Because they are highly aggregated (NRC, 2007c), it was not possible to identify the successful programs that are nearing completion that could be replaced by new research initiatives.

Substantial new investment is required to implement the other major initiatives proposed in this report, including regional modeling, a climate observing system, and climate services. A small fraction of the required funding may be found through budget trades (e.g., redirect some funding from global modeling to regional modeling) or partnerships with relevant international, state, regional, and local efforts and/or with federal agencies that have had little or no participation in the CCSP. New partnerships with the intelligence community, for example, may result in new funding for research on climate impacts, which are relevant to a wide variety of issues including national security. However, significant funds for implementing the major initiatives cannot be diverted from the current program, which provides the underpinning research and must be sustained.

Management Challenges

Implementing the priorities identified by the committee will not be easy. The program faces a number of management challenges, including an interagency structure, insufficient attention from key White House offices, a natural science culture, inadequate community capacity in critical areas, and a broad mandate that requires coordination with other interagency programs. Although the committee offers a few suggestions about how to overcome these management challenges, it had neither the charge nor expertise to evaluate different program structures (e.g., a single climate agency versus interagency coordination) or to prescribe how an interagency program should operate.

The interagency structure is both a strength and a weakness of the program. The CCSP coordinates the climate change programs of 13 agencies, each of which designates a piece of its program portfolio as part of the CCSP. The major strength of this approach is its potential to harness the expertise and funding needed to carry out program goals and objectives. Weaknesses include the following:

• CCSP priorities usually do not align directly with agency and department priorities, making it difficult to match agency and CCSP programs and thus to obtain funding for CCSP priorities.
• The need for multiple agencies to coordinate activities poses a high administrative burden in the form of additional meetings and reporting. This burden is increased by the need for the same agencies to coordinate activities with related programs, such as a national climate service (if developed outside the program), the CCTP, the Subcommittee on Ocean Science and Technology, and international research and observing programs.

These problems would be exacerbated in the climate change research program envisioned in this report because more federal agencies as well as state and local government agencies and emerging potential partners (e.g., nongovernmental organizations, foundations, businesses) would be involved. However, the problems are not insurmountable, even in an interagency structure. The most important factor is good leaders with the authority to direct resources and/or research to achieve program goals (NRC, 2005b). A charismatic leader with strong scientific credentials is also needed to communicate the importance of taking an end-to-end approach to the climate problem and convincing agency heads and appropriators to make the necessary investments.

Although the CCSP has a director (an acting director for several years), he has authority to direct only that part of the program funded through his agency. The managers responsible for implementation have even less authority over budgets and programs. The absence of centralized budget authority limits the ability of the

CCSP to influence the climate priorities of participating agencies or implement new research directions that fall outside or across agency missions (NRC, 2007c). An increased discretionary budget for the CCSP director, sufficient to carry out interagency efforts such as workshops and tactical studies, would provide flexibility and seed money for objectives that are of higher priority to the program than to any participating agency.

Another principle for successful organizations is that what gets attention gets done.[13] In the early years of the predecessor USGCRP, a close working relationship between the Office of Management and Budget and agency leaders was instrumental in securing funding for key program priorities (NRC, 1999b). A similar relationship is even more important now, given the large number of congressional committees responsible for appropriating climate research funding. However, even a management structure intended to provide cabinet-level oversight of the CCSP (and the CCTP) has not resulted in strong linkages between the CCSP, CCTP, and the White House. The creation of a climate czar position and the appointments of respected scientists with interests in climate and energy to lead OSTP, DOE, and NOAA in the new administration should provide the level of attention needed to make the program succeed. It should also strengthen coordination of climate change science and technologies across the federal government. Such high-level attention is especially important for observations, which underpin the entire research program, but are chosen primarily by NASA (satellite observations take up nearly half of the CCSP budget) and other individual agencies. Reconciling the different priorities and planning horizons is essential for developing the knowledge foundation needed to address climate-related problems.

Another leadership issue concerns the human dimensions of climate change. The relevant programs are small compared to natural science programs and scattered around different agencies. This makes it difficult for human dimensions program managers to take a strong leadership role in the CCSP, which in turn makes it difficult to move the CCSP in new directions. The result is that

---

[13] Presentation to the committee by Robert Waterman, management consultant, The Waterman Group, Inc., on March 21, 2008.

even with 13 agencies participating in the program, CCSP agency leaders have relatively little expertise in the human dimensions of climate change or in adaptation and mitigation research. It seems unlikely that the future climate change research program will be able to take a more comprehensive view of the climate–human–environmental system unless an agency devoted to basic and applied social science research (such as NSF) steps up to help organize and build the research community and bring a stronger human dimensions perspective to the program leadership. For example, a strong human dimensions program leader would be able to work with natural science counterparts to develop integrated research teams to work together on the scientific-societal issues outlined in Chapter 2.

Building the human dimensions research community will be important not only for the research component of the future climate change research program, but also for climate services and a national assessment of climate impacts and adaptation options. The latter has the potential to overburden a small community that is already participating in the IPCC assessments. Indeed, the much larger natural science community is struggling to contribute to these assessments while continuing to generate new research results. Because national and international assessments are valuable for monitoring climate change and impacts and for summarizing what is known for policy makers, the future climate change research program will have to take steps to minimize the burden on the scientific community. Approaches that might be taken include limiting the scope of ongoing assessments to significant new developments and timing new assessments to optimize the ability to build on previous assessments (NRC, 2007a).

Finally, the increased demand for climate information has amplified the importance of providing information that users can trust. Examples of political considerations dictating what climate research results are communicated have been widely reported (e.g., Donaghy et al., 2007; House of Representatives, 2007). Even the possibility that research results have been withheld, delayed, or selectively interpreted can weaken trust in the program and discourage decision makers from using science-based information. The most effective way to guard against political interference is to institute transparent processes for key stages of research, from selecting priorities and approaches to peer reviewing scientific results, and to give a restructured climate change science program the authority to communicate results to the public in a timely fashion.

Climate change is critically important to our nation and the world. Addressing the challenges posed by climate change will require a strengthened research program aimed at understanding climate variability and change as well as supporting robust approaches for mitigating the causes and anticipating and adapting to the expected changes. Although this end-to-end approach was called for in the CCSP strategic plan (CCSP, 2003), for it to be realized, the emphasis will have to be shifted toward understanding the complex interactions between climate, humans, and the environment. This, in turn, will require a more integrated approach to research—one without the false dichotomies between natural and social science, between scientific disciplines, and between basic and applied science. To ensure that this shift also succeeds in producing information that decision makers need, stronger connections will have to be forged with major groups of stakeholders (e.g., water resource and land managers, policy makers), who can contribute data to support research objectives, guide the development of a national assessment and a national climate service, and benefit from the results. Fortunately, the successes of the CCSP and its predecessor USGCRP provide a strong foundation for making this transition to meet today's challenges.

References

Adger, N.W., S. Huq, K. Brown, D. Conway, and M. Hulme, 2003, Adaptation to climate change in the developing world, *Progress in Development Studies*, **3**, 179–195.

Adger, N., T. Hughes, C. Folke, S. Carpenter, and J. Rockstrom, 2005, Socio-ecological resilience to coastal disasters, *Science*, **309**, 1036–1039.

Adger, W.N., S. Agrawala, M.M.Q. Mirza, C. Conde, K. O'Brien, J. Pulhin, R. Pulwarty, B. Smit, and K. Takahashi, 2007, Assessment of adaptation practices, options, constraints and capacity, in *Climate Change 2007: Impacts, Adaptation and Vulnerability*, Contribution of Working Group II to the Fourth Assessment Report of the Intergovernmental Panel on Climate Change, M.L. Parry, O.F. Canziani, J.P. Palutikof, P.J. van der Linden, and C.E. Hanson, eds., Cambridge University Press, Cambridge, pp. 717–743.

Agrawal, A., 2008, The role of local institutions in adaptation to climate change, in *The Social Dimensions of Climate Change*, Social Development Department, The World Bank, Washington, D.C., 65 pp.

Alexander, L.V., X. Zhang, T.C. Peterson, J. Caesar, B. Gleason, A.M.G. Klein Tank, M. Haylock, D. Collins, B. Trewin, F. Rahimzadeh, A. Tagipour, K. Rupa Kumar, J. Revadekar, G. Griffiths, L. Vincent, D.B. Stephenson, J. Burn, E. Aguilar, M. Brunet, M. Taylor, M. New, P. Zhai, M. Rusticucci, and J.L. Vazquez-Aguirre, 2006, Global observed changes in daily climate extremes of temperature and precipitation, *Journal of Geophysical Research*, **111**, D05109, doi:10.1029/2005JD006290.

Alley, R.B., P.U. Clark, P. Huybrechts, and I. Joughin, 2005, Ice-sheet and sea-level changes, *Science*, **310**, 456–460.

Auffhammer, M., and R.T. Carson, 2008, Forecasting the path of China's CO_2 emissions using province level information, *Journal of Environmental Economics and Management*, **55**, 229–247.

Backlund, P., A. Janetos, D. Schimel, J. Hatfield, K. Boote, P. Fay, L. Hahn, C. Izaurralde, B.A. Kimball, T. Mader, J. Morgan, D. Ort, W. Polley, A. Thomson, D. Wolfe, M.G. Ryan, S.R. Archer, R. Birdsey, C. Dahm, L. Heath, J. Hicke, D. Hollinger, T. Huxman, G. Okin, R. Oren, J. Randerson, W. Schlesinger, D. Lettenmaier, D. Major, L. Poff, S. Running, L. Hansen, D. Inouye, B.P. Kelley, L. Meyerson, B. Paterson, and R. Shaw, eds., 2008, *The Effects of Climate Change on Agriculture, Land Resources, Water Resources, and Biodiversity in the United States*, Synthesis and Assessment Product 4.3, Climate Change Science Program and Subcommittee on Global Change Research, Washington, D.C., 362 pp.

Bala, G., K. Caldeira, M. Wickett, T.J. Phillips, D.B. Lobell, C. Delire, and A. Mirin, 2007, Combined climate and carbon-cycle effects of large-scale deforestation, *Proceedings of the National Academy of Sciences*, **104**, 6550–6555.

Bates, B.C., Z.W. Kundzewicz, S. Wu, and J.P. Palutikof, eds., 2008, *Climate Change and Water*, Technical Paper of the Intergovernmental Panel on Climate Change, IPCC Secretariat, Geneva, 210 pp.

Batterbury, S., and A. Warren, 2001, The African Sahel 25 years after the great drought: Assessing progress and moving towards new agendas and approaches, *Global Environmental Change*, **11**, 1–8.

Beller-Simms, N., H. Ingram, D. Feldman, N. Mantua, K.L. Jacobs, and A. Waple, eds., 2008, *Decision-Support Experiments and Evaluations Using Seasonal to Interannual Forecasts and Observational Data: A Focus on Water Resources*, Synthesis and Assessment Product 5.3, Climate Change Science Program and Subcommittee on Global Change Research, Washington, D.C., 190 pp.

Bonan, G.B., 2008, Forests and climate change: Forcings, feedbacks, and the climate benefits of forests, *Science*, **320**, 1444–1449.

Bozmoski, A., M.C. Lemos, and E. Young, 2008, Prosperous negligence: Governing the clean development mechanism for markets and development, *Environment*, **50**, 18.

Breshears, D.D., N.S. Cobb, P.M. Rich, K.P. Price, C.D. Allen, R.G. Balice, W.H. Romme, J.H. Kastens, M.L. Floyd, J. Belnap, J.J. Anderson, O.B. Myers, and C.W. Meyer, 2005, Regional vegetation die-off in response to global-change-type drought, *Proceedings of the National Academy of Sciences*, **102**, 15,144–15,148.

Buddemeier, R.W., J.A. Kleypas, and R.B. Aronson, 2004, *Coral Reefs and Global Climate Change: Potential Contributions of Climate Change to Stresses on Coral Reefs*, Pew Center on Global Climate Change, 35 pp., available at *http://www.pewclimate.org/global-warming-in-depth/all_reports/coral_reefs*.

Buesseler, K.O., S.C. Doney, D.M. Karl, P.W. Boyd, K. Caldeira, F. Chai, K.H. Coale, H.J.W. de Baar, P.G. Falkowski, K.S. Johnson, R.S. Lampitt, A.F. Michaels, S.W.A. Naqvi, V. Smetacek, S. Takeda, and A.J. Watson, 2008, Ocean iron fertilization—Moving forward in a sea of uncertainty, *Science*, **319**, 162.

Burby, R.J., 2006, Hurricane Katrina and the paradoxes of government disaster policy: Bringing about wise governmental decisions for hazardous areas, *Annals of the American Academy of Political and Social Science*, **604**, 171–191.

Burkett, V.R., D.A. Wilcox, R. Stottlemyer, W. Barrow, D. Fagre, J. Baron, J. Price, J.L. Nielsen, C.D. Allen, D.L. Peterson, G. Ruggerone, and T. Doyle, 2005, Nonlinear dynamics in ecosystem response to climatic change: Case studies and policy implications, *Ecological Complexity*, **2**, 357–394.

Busby, J.W., 2007, *Climate Change and National Security: An Agenda for Action*, Council Special Report 32, Council on Foreign Relations, Brookings Institution Press, New York, 32 pp.

CARB (California Air Resources Board), 2008, *Climate Change Proposed Scoping Plan: A Framework for Change*, 122 pp., available at *http://www.arb.ca.gov/cc/scopingplan/document/psp.pdf*.

Carpenter, K.E., M. Abrar, G. Aeby, R.B. Aronson, S. Banks, A. Bruckner, A. Chiriboga, J. Cortés, J.C. Delbeek, L. DeVantier, G.J. Edgar, A.J. Edwards, D. Fenner, H.M. Guzmán, B.W. Hoeksema, G. Hodgson, O. Johan, W.Y. Licuanan, S.R. Livingstone, E.R. Lovell, J.A. Moore, D.O. Obura, D. Ochavillo, B.A. Polidoro, W.F. Precht, M.C. Quibilan, C. Reboton, Z.T. Richards, A.D. Rogers, J. Sanciangco, A. Sheppard, C. Sheppard, J. Smith, S. Stuart, E. Turak, J.E.N. Veron, C. Wallace, E. Weil, and E. Wood, 2008, One-third of reef-building corals face elevated extinction risk from climate change and local impacts, *Science*, **321**, 560–563.

Cazenave, A., and R.S. Nerem, 2004, Present-day sea level change: Observations and causes, *Reviews of Geophysics*, **42**, RG3001, doi:10.1029/2003RG000139.

CCSP (Climate Change Science Program), 2003, *Strategic Plan for the U.S. Climate Change Science Program*, Climate Change Science Program and Subcommittee on Global Change Research, Washington, D.C., 202 pp.

CCSP, 2008, *Our Changing Planet: The U.S. Climate Change Science Program for Fiscal Year 2008*, Climate Change Science Program and Subcommittee on Global Change Research, Washington, D.C., 212 pp.

CCSP, 2009, *Coastal Sensitivity to Sea Level Rise: A Focus on the Mid-Atlantic Region*, Synthesis and Assessment Product 4.1, Climate Change Science Program and Subcommittee on Global Change Research, Washington, D.C., 784 pp.

CCTP (Climate Change Technology Program), 2006, *U.S. Climate Change Technology Program Strategic Plan*, DOE/PI-0005, Washington, D.C., 224 pp., available at *http://www.climatetechnology.gov/stratplan/final/index.htm*.

CDWR (California Department of Water Resources), 2007, *Proceedings of the Western Governors' Association, Western States Water Council, and California Department of Water Resources May 2007 Climate Change Research Needs Workshop*, Sacramento, CA, 51 pp., available at *http://wwwdwr.water.ca.gov/climatechange/docs/ClimateChangeReport-100307.pdf*.

CDWR, 2008a, *California Drought, an Update, 2008*, Sacramento, CA, 59 pp.

CDWR, 2008b, *Managing an Uncertain Future: Adaptation Strategies for Climate Change and California's Water*, Sacramento, CA, 30 pp., available at *http://www.water.ca.gov/climatechange/docs/ClimateChangeWhitePaper.pdf*.

CENR (Committee on Environment and Natural Resources), 2008, *Scientific Assessment of the Effects of Global Change on the United States*, Climate Change Science Program, Washington, D.C., 261 pp.

CEOS (Committee on Earth Observation Satellites), 2006, *Satellite Observation of the Climate System*, CEOS Response to the GCOS Implementation Plan, 54 pp., available at *http://www.ceos.org/pages/CEOSResponse_1010A.pdf*.

Cesar, H., P. van Beukering, S. Pintz, and J. Dierking, 2002, *Economic Valuation of the Coral Reefs of Hawaii*, Final report to National Oceanic and Atmospheric Administration Coastal Ocean Program, Cesar Environmental Economics Consulting, Arnhem, The Netherlands, 123 pp.

Cesar, H., L. Burke, and L. Pet-Soede, 2003, *The Economics of Worldwide Coral Reef Degradation*, Cesar Environmental Economics Consulting (CEEC), Arnhem, The Netherlands, 23 pp.

Challinor A., T. Wheeler, C. Garforth, P. Craufurd, and A. Kassam, 2007, Assessing the vulnerability of food crop systems in Africa to climate change, *Climate Change*, **83**, 381–399.

Chambers, J.Q., J.I. Fisher, H. Zeng, E.L. Chapman, D.B. Baker, and G.C. Hurtt, 2007, Hurricane Katrina's carbon footprint on U.S. Gulf Coast forests, *Science*, **318**, 1107.

Chan, F., J.A. Barth, J. Lubchenco, A. Kirincich, H. Weeks, W.T. Peterson, and B.A. Menge, 2008, Emergence of anoxia in the California Current large marine ecosystem, *Science*, **319**, 920.

Christopolos, I., 2008, *Incentives and Constraints to Climate Change Adaption and Disaster Risk Reduction – a Local Perspective*, Commission on Climate Change and Development, 9 pp., available at *http://www.ccdcommission.org/Filer/pdf/pb_incentives_linking_ climate_change.pdf*.

Church, J.A., and N.J. White, 2006, A 20th century acceleration in global sea-level rise, *Geophysical Research Letters*, **33**, L01602, doi:10.1029/2005GL024826.

CNA Corporation, 2007, *National Security and the Threat of Climate Change*, Alexandria, VA, 63 pp.

Cook, T., M. Folli, J. Klinck, S. Ford, and J. Miller, 1998, The relationship between increasing sea-surface temperature and the northward spread of *Perkinsus marinus* (Dermo) disease epizootics in oysters, *Estuarine and Coastal Shelf Science*, **46**, 587–597.

Cutter, S.L., and C.T. Emrich, 2006, Moral hazard, social catastrophe: The changing face of vulnerability along the hurricane coasts, *Annals of the American Academy of Political and Social Science*, **604**, 102–112.

Dai, A., 2006, Recent climatology, variability, and trends in global surface humidity, *Journal of Climate*, **19**, 3589–3606.

Dai, A.G., K.E. Trenberth, and T.T. Qian, 2004, A global dataset of Palmer Drought Severity Index for 1870–2002: Relationship with soil moisture and effects of surface warming, *Journal of Hydrometeorology*, **5**, 1117–1130.

Das, S.B., I. Joughin, M.D. Behn, I.M. Howat, M.A. King, D. Lizarralde, and M.P. Bhatia, 2008, Fracture propagation to the base of the Greenland Ice Sheet during supraglacial lake drainage, *Science*, **320**, 778–781.

Del Toro-Silva, F.M., J.M. Miller, J.C. Taylor, and T.A. Ellis, 2008, Influence of oxygen and temperature on growth and metabolic performance of *Paralichthys lethostigma* (Pleuronectiformes: Paralichthyidae), *Journal of Experimental Marine Biology and Ecology*, **358**, 113–123.

Déry, S.J., and R.D. Brown, 2007, Recent Northern Hemisphere snow cover extent trends and implications for the snow-albedo-feedback, *Geophysical Research Letters*, **34**, L22504, doi:10.1029/2007GL031474.

Diaz, R.J., and R. Rosenberg, 2008, Spreading dead zones and consequences for marine ecosystems, *Science*, **321**, 926–929.

Dietz, S., and N. Stern, 2008, Why economic analysis supports strong action on climate change: A response to the Stern Review's critics, *Review of Environmental Economics and Policy*, **2**, 94–113.

Dilling, L., S.C. Doney, J. Edmonds, K.R. Gurney, R. Harriss, D. Schimel, B. Stephens, and G. Stokes, 2003, The role of carbon cycle observations and knowledge in carbon management, *Annual Review of Environment and Resources*, **28**, 521–558.

DOC (Department of Commerce), 2005, Billion dollar U.S. weather disasters, National Oceanic and Atmospheric Administration, National Climatic Data Center, available at *http://www.ncdc.noaa.gov/oa/reports/billionz.html*.

DOE (Department of Energy), 2008, *Identifying Outstanding Grand Challenges in Climate Change Research: Guiding DOE's Strategic Planning*, Report on the DOE/BERAC Workshop, March 25–27, 2008, Arlington, VA, 38 pp., available at *http://www.sc.doe.gov/ober/berac/Grand_Challenges_Report.pdf*.

Donaghy, T., J. Freeman, F. Grifo, K. Kaufman, T. Maassarani, and L. Shultz, 2007, *Atmosphere of Pressure: Political Interference in Federal Climate Science*, Union of Concerned Scientists and the Governmental Accountability Project, Cambridge, MA, 80 pp., available at *http://www.ucsusa.org/assets/documents/scientific_integrity/atmosphere-of-pressure.pdf*.

Eakin, H., and M.C. Lemos, 2006, Adaptation and the state: Latin America and the challenge of capacity-building under globalization, *Global Environmental Change*, **16**, 7–18.

Eakin, H., and A.L. Luers, 2006, Assessing the vulnerability of social-environmental systems, *Annual Review of Environment and Resources*, **31**, 365–394.

Easterling, D.R., D.M. Anderson, S.J. Cohen, W.J. Gutowski, Jr., G.J. Holland, K.E. Kunkel, T.C. Peterson, R.S. Pulwarty, R.J. Stouffer, and M.F. Wehner, 2008, Measures to improve our understanding of weather and climate extremes, in *Weather and Climate Extremes in a Changing Climate. Regions of Focus: North America, Hawaii, Caribbean, and U.S. Pacific Islands*, T.R. Karl, G.A. Meehl, C.D. Miller, S.J. Hassol, A.M. Waple, and W.L. Murray, eds., Synthesis and Assessment Product 3.3, Climate Change Science Program and Subcommittee on Global Change Research, Washington, D.C., pp. 117–126.

Easterling, W.E., P.K. Aggarwal, P. Batima, K.M. Brander, L. Erda, S.M. Howden, A. Kirilenko, J. Morton, J.-F. Soussana, J. Schmidhuber, and F.N. Tubiello, 2007, Food, fibre and forest products, in *Climate Change 2007: Impacts, Adaptation and Vulnerability*, Contribution of Working Group II to the Fourth Assessment Report of the Intergovernmental Panel on Climate Change, M.L. Parry, O.F. Canziani, J.P. Palutikof, P.J. van der Linden, and C.E. Hanson, eds., Cambridge University Press, Cambridge, pp. 273–313.

Edmonds, J., and J. Reilly, 1983, A long-term global energy-economic model of carbon-dioxide release from fossil-fuel use, *Energy Economics*, **5**, 74–88.

EIA (Energy Information Administration), 2008a, *Energy Market and Economic Impacts of S. 2191, the Lieberman-Warner Climate Security Act of 2007*, SR/OIAF/2008-01, Department of Energy, Washington, D.C., 60 pp., available at *http://www.eia.doe.gov/oiaf/servicerpt/s2191/index.html*.

EIA, 2008b, *International Energy Outlook 2008*, DOE/EIA-0484(2008), Department of Energy, Washington, D.C., 250 pp., available at *http://www.eia.doe.gov/oiaf/ieo/*.

Ellerman, A.D., P.L. Joskow, R. Schmalensee, J.P. Montero, and E.M. Bailey, 2000, *Markets for Clean Air: The U.S. Acid Rain Program*, Cambridge University Press, Cambridge, 362 pp.

Elsner, J.B., J.P. Kossin, and T.H. Jagger, 2008, The increasing intensity of the strongest tropical cyclones, *Nature*, **455**, 92–95.

Emanuel, K.A., 2005, Increasing destructiveness of tropical cyclones over the past 30 years, *Nature*, **436**, 686–688.

Emanuel, K.A., 2007, Environmental factors affecting tropical cyclone power dissipation, *Journal of Climate*, **20**, 5497–5509.

ESSC (Earth System Sciences Committee), 1988, *Earth System Science: A Closer View*, National Aeronautics and Space Administration, Washington, D.C., 208 pp. plus annexes.

Ezzati, M., A. Lopez, A. Rodgers, and C. Murray, eds., 2004, *Comparative Quantification of Health Risks: Global and Regional Burden of Disease Due to Selected Major Risk Factors*, World Health Organization, Geneva, 2248 pp.

FAO (Food and Agriculture Organization), 2008, *Climate Change and Food Security: A Framework Document*, Rome, 93 pp.

Feldman, D.L., 2007, *Water Policy for Sustainable Development*, Johns Hopkins University Press, Baltimore, MD, 371 pp.

Fischer, C., and R.D. Morgenstern, 2006, Carbon abatement costs: Why the wide range of estimates? *Energy Journal*, **27**, 73–86.

Folke, C., 2006, Resilience: The emergence of a perspective for social-ecological systems analyses, *Global Environmental Change*, **16**, 253–267.

Freeman, J., and C. Kolstad, eds., 2006, *Moving to Markets in Environmental Regulation*, Oxford University Press, New York, 500 pp.

Gamble, J.L., ed., 2008, *Analyses of the Effects of Global Change on Human Health and Welfare and Human Systems*, Synthesis and Assessment Product 4.6, Climate Change Science Program and Sub-committee on Global Change Research, Washington, D.C., 271 pp.

GCOS (Global Climate Observing System), 2003, *The Second Report on the Adequacy of the Global Observing Systems for Climate in Support of the UNFCCC*, GCOS 82, WMO/TD 1143, World Meteorological Organization, Geneva, 74 pp., available at *http://www.wmo.ch/pages/ prog/gcos/Publications/gcos-82_2AR.pdf.*

GCOS, 2004, *Implementation Plan for the Global Observing System for Climate in Support of the UNFCCC*, GCOS 92, WMO/TD 1219, World Meteorological Organization, Geneva, 136 pp., available at *http://www.wmo.int/pages/prog/gcos/Publications/gcos-92_GIP.pdf.*

GCOS, 2006, *Systematic Observation Requirements for Satellite-Based Products for Climate*, GCOS-107, WMO/TD 1338, Geneva, 90 pp.

Gerber, B.J., 2007, Disaster management in the United States: Examining key political and policy challenges, *Policy Studies Journal*, **35**, 227–238.

Glantz, M., and D. Jamieson, 2000, Societal response to Hurricane Mitch and intra- versus intergenerational equity issues: Whose norms should apply? *Risk Analysis*, **20**, 869–882.

Global Carbon Project, 2008, *Carbon Budget and Trends 2007*, available at *http://www.globalcarbonproject.org.*

Graham, L.P., S. Hagemann, S. Juan, and M. Beniston, 2007, On inter-preting hydrological change from regional climate models, *Climatic Change*, **81**, 97–122.

Groisman, P.Y., R.W. Knight, T.R. Karl, D.R. Easterling, B. Sun, and J.H. Lawrimore, 2004, Contemporary changes of the hydrological cycle over the contiguous United States: Trends derived from in situ observations, *Journal of Hydrometeorology*, **5**, 64–85.

Gulev, S.K., O. Zolina, and S. Grigoriev, 2001, Extratropical cyclone vari-ability in the Northern Hemisphere winter from the NCEP/NCAR reanalysis data, *Climate Dynamics*, **17**, 795–809.

Gutowski, W.J., G.C. Hegerl, G.J. Holland, T.R. Knutson, L.O. Mearns, R.J. Stouffer, P.J. Webster, M.F. Wehner, and F.W. Zwiers, 2008, Causes of observed changes in extremes and projections of future changes, in *Weather and Climate Extremes in a Changing Climate. Regions of Focus: North America, Hawaii, Caribbean, and U.S. Pa-cific Islands*, Synthesis and Assessment Product 3.3, Climate Change Science Program and Subcommittee on Global Change Research, Washington, D.C., pp. 81–116.

Haberl, H., K.H. Erb, F. Krausmann, V. Gaube, A. Bondeau, C. Plutzar, S. Gingrich, W. Lucht, and M. Fischer-Kowalski, 2007, Quantifying and mapping the human appropriation of net primary production in

Earth's terrestrial ecosystems, *Proceedings of the National Academy of Sciences*, **104**, 12,942–12,945.

Hanemann, M., 2008, California's new greenhouse gas laws, *Review of Environmental Economics and Policy*, **2**, 114–129.

Harrison, P.J., F.A. Whitney, D.L. Mackas, R.J. Beamish, M. Trudel, and I.R. Perry, 2005, Changes in coastal ecosystems in the NE Pacific Ocean, in *Proceedings of the International Symposium on Long-Term Variations in the Coastal Environments and Ecosystems*, September 27–28, 2005, Matsuyama, Japan, pp. 17–35.

Harvell, C.D., K. Kim, J.M. Burkholder, R.R. Colwell, P.R. Epstein, D.J. Grimes, E.E. Hofmann, E.K. Lipp, A.D.M.E. Osterhaus, R.M. Overstreet, J.W. Porter, G.W. Smith, and G.R. Vasta, 1999, Emerging marine diseases—Climate links and anthropogenic factors, *Science*, **285**, 1505–1510.

Harvell, D., C.E. Mitchell, J. Ward, S. Altizer, A.P. Dobson, R.S. Ostfeld, and M.D. Samuel, 2002, Climate warming and disease risks for terrestrial and marine biota, *Science*, **296**, 2158–2162.

Harvell, D., R.B. Aronson, N. Baron, J.H. Connell, A.P. Dobson, S.P. Ellner, L. Gerber, K. Kim, A. Kuris, H. McCallum, K. Lafferty, B. McKay, J.W. Porter, M. Pascual, G. Smith, K. Sutherland, and J. Ward, 2004, The rising tide of ocean diseases: Unsolved problems and research priorities, *Frontiers in Ecology and the Environment*, **7**, 375–382.

Hassol, S.J., 2004, *Impacts of a Warming Arctic: Arctic Climate Impact Assessment*, Cambridge University Press, New York, 139 pp.

Hatfield, J., K. Boote, P. Fay, L. Hahn, C. Izaurralde, B.A. Kimball, T. Mader, J. Morgan, D. Ort, W. Polley, A. Thomson, and D. Wolfe, 2008, Agriculture, in *The Effects of Climate Change on Agriculture, Land Resources, Water Resources, and Biodiversity in the United States*, Synthesis and Assessment Product 4.3, Climate Change Science Program and Subcommittee on Global Change Research, Washington, D.C., pp. 21–74.

Heal, G., 2009, Climate economics: A meta-review and some suggestions for future research, *Review of Environmental Economics and Policy*, **3**, 4–21.

Hoerling, M., J. Hurrell, J. Eischeid, and A. Phillips, 2006, Detection and attribution of twentieth-century northern and southern African rainfall change, *Journal of Climate*, **19**, 3989–4008.

Hohenemser, C., R.E. Kasperson, and R.W. Kates, 1985, Causal structure, in *Perilous Progress: Managing the Hazards of Technology*, R.W. Kates, C. Hohenemser, and J.X. Kasperson, eds., Westview Press, Boulder, CO, pp. 45–66.

Holland, D.M., R.H. Thomas, B. de Young, M.H. Ribergaard, and B. Lyberth, 2008, Acceleration of Jakobshavn Isbrae triggered by warm subsurface ocean waters, *Nature Geoscience*, **1**, 659–664.

Holland, G.J., and P.J. Webster, 2007, Heightened tropical cyclone activity in the North Atlantic: Natural variability or climate trend? *Philosophical Transactions of the Royal Society A*, **365**, 2695–2716.

House of Representatives, 2007, *Political Interference with Climate Change Science Under the Bush Administration*, Committee on Oversight and Government Reform, 33 pp., available at *http://oversight.house.gov/documents/20071210101633.pdf*.

Howat, I.M., I. Joughin, and T. Scambos, 2007, Rapid changes in ice discharge from Greenland outlet glaciers, *Science*, **315**, 1559–1561.

Howden, S.M., J.F. Soussana, F. Tubiello, N. Chhetri, M. Dunlop, and H. Meinke, 2007, Adapting agriculture to climate change, *Proceedings of the National Academy of Sciences*, **104**, 19,691–19,696.

IEA (International Energy Agency), 2007, *World Energy Outlook 2007: China and India Insights*, International Energy Agency, Paris, 670 pp., available at *http://www.worldenergyoutlook.org/ 2007.asp*.

Iglesias-Rodriguez, M.D., P.R. Halloran, R.E.M. Rickaby, I.R. Hall, E. Colmenero-Hidalgo, J.R. Gittins, D.R.H. Green, T. Tyrrell, S.J. Gibbs, P. von Dassow, E. Rehm, E.V. Armbrust, and K.P. Boessenkool, 2008, Phytoplankton calcification in a high-CO_2 world, *Science*, **320**, 336–340.

Ingram, J.S.I., P.J. Gregory, and A-M. Izac, 2008, The role of agronomic research in climate change and food security policy, *Agriculture, Ecosystems and Environment*, **126**, 4–12.

IPCC (Intergovernmental Panel on Climate Change), 1995, *Climate Change 1995: Economic and Social Dimensions of Climate Change*, Contribution of Working Group III to the Second Assessment of the Intergovernmental Panel on Climate Change, J.P. Bruce, H. Lee, and E.F. Haites, eds., Cambridge University Press, Cambridge, 448 pp.

IPCC, 2001a, *Climate Change 2001: Synthesis Report*, Contribution of Working Groups I, II, and III to the Third Assessment Report of the Intergovernmental Panel on Climate Change, R.T. Watson and the Core Writing Team, eds., Cambridge University Press, Cambridge, 398 pp.

IPCC, 2001b, *Climate Change 2001: The Scientific Basis*, Contribution of Working Group I to the Third Assessment Report of the Intergovernmental Panel on Climate Change, J.T. Houghton, Y. Ding, D.J. Griggs, M. Noguer, P. J. van der Linden, and D. Xiaosu, eds., Cambridge University Press, Cambridge, 944 pp.

IPCC, 2007a, *Climate Change 2007: Synthesis Report*, Contribution of Working Groups I, II and III to the Fourth Assessment Report of the

Intergovernmental Panel on Climate Change, Core Writing Team, R.K. Pachauri, and A. Reisinger, eds., Geneva, 104 pp.

IPCC, 2007b, *Climate Change 2007: The Physical Science Basis*, Contribution of Working Group I to the Fourth Assessment Report of the IPCC, S. Solomon, D. Qin, M. Manning, Z. Chen, M. Marquis, K.B. Averyt, M. Tignor and H.L. Miller, eds., Cambridge University Press, Cambridge, 996 pp.

IPCC, 2007c, *Climate Change 2007: Impacts, Adaptation and Vulnerability*, Contribution of Working Group II to the Fourth Assessment Report of the Intergovernmental Panel on Climate Change, M.L. Parry, O.F. Canziani, J.P. Palutikof, P.J. van der Linden, and C.E. Hanson, eds., Cambridge University Press, Cambridge, 976 pp.

IPCC, 2007d, *Climate Change 2007: Mitigation*, Contribution of Working Group III to the Fourth Assessment Report of the Intergovernmental Panel on Climate Change, B. Metz, O.R. Davidson, P.R. Bosch, R. Dave, and L.A. Meyer, eds., Cambridge University Press, Cambridge, 851 pp.

Ivey, J.L., J. Smithers, R.C. de Loë, and R.D. Kreutzwiser, 2004, Community capacity for adaptation to climate-induced water shortages: Linking institutional complexity and local actors, *Environmental Management*, **33**, 36–47.

Jackson, R.B., E.G. Jobbágy, R. Avissar, S.B. Roy, D.J. Barrett, C.W. Cook, K.A. Farley, D.C. le Maitre, B.A. McCarl, and B.C. Murray, 2005, Trading water for carbon with biological sequestration, *Science*, **310**, 1944–1947.

Johns, G.M., V.R. Leeworthy, F.W. Bell, and M.A. Bonn, 2001, *Socioeconomic Study of Reefs in Southeast Florida*, Final report submitted to Broward County, Palm Beach County, Miami-Dade County, Monroe County, Florida Fish and Wildlife Conservation Commission, and National Oceanic and Atmospheric Administration, as revised April 18, 2003, 348 pp.

Joughin, I., S.B. Das, M.A. King, B.E. Smith, I.M. Howat, and T. Moon, 2008, Seasonal speedup along the western flank of the Greenland Ice Sheet, *Science*, **320**, 781–783.

Julius, S.H., J.M. West (eds.), J.S. Baron, B. Griffith, L.A. Joyce, P. Kareiva, B.D. Keller, M.A. Palmer, C.H. Peterson, and J.M. Scott (authors), 2008, *Preliminary Review of Adaptation Options for Climate-Sensitive Ecosystems and Resources*, Synthesis and Assessment Product 4.4, Climate Change Science Program and Subcommittee on Global Change Research, Washington, D.C., 873 pp.

Justic, D., N.N. Rabalais, and R.E. Turner, 1997, Impacts of climate change on net productivity of coastal waters: Implications for carbon budgets and hypoxia, *Climate Change*, **8**, 225–237.

Kalnay, E., M. Kanamitsu, R. Kistler, W. Collins, D. Deaven, L. Gandin, M. Iredell, S. Saha, G. White, J. Woollen, Y. Zhu, A. Leetmaa, B. Reynolds, M. Chelliah, W. Ebisuzaki, W. Higgins, J. Janowiak, K.C. Mo, .C. Ropelewski, J. Wang, R. Jenne, and D. Joseph, 1996, The NCEP/NCAR 40-year reanalysis project, *Bulletin of the American Meteorological Society*, **77**, 437–471.

Karl, T.R., G.A. Meehl, C.D. Miller, S.J. Hassol, A.M. Waple, and W.L. Murray, eds., 2008, *Weather and Climate Extremes in a Changing Climate. Regions of Focus: North America, Hawaii, Caribbean, and U.S. Pacific Islands*, Synthesis and Assessment Product 3.3, Climate Change Science Program and Subcommittee on Global Change Research, Washington, D.C., 164 pp.

Kates, R., 2000, Population and consumption: What we know, what we need to know, *Environment*, **10**, 12–19.

Keller, M., D.S. Schimel, W.W. Hargrove, and F.M. Hoffman, 2008, A continental strategy for the National Ecological Observatory Network, *Frontiers in Ecology and the Environment*, **6**, 282–284.

Kelly, D.L., C.D. Kolstad, and G.T. Mitchell, 2005, Adjustment costs from environmental change, *Journal of Environmental Economics and Management*, **50**, 468–495.

Klein Tank, A.M.G., and G.P. Können, 2003, Trends in indices of daily temperature and precipitation extremes in Europe, 1946–1999, *Journal of Climate*, **16**, 3665–3680.

Kolstad, C.D., and M. Toman, 2005, The economics of climate policy, in *Handbook of Environmental Economics*, K.-G. Maler and J. Vincent, eds., Vol. 3, Elsevier, Amsterdam, pp. 1561–1618.

Kleinosky, L.R., B. Yarnal, and A. Fisher, 2007, Vulnerability of Hampton Roads, Virginia to storm-surge flood and sea-level rise, *Natural Hazards*, **40**, 43–70.

Kleypas J.A., R.A. Feely, V.J. Fabry, C. Langdon, C.L. Sabine, and L.L. Robbins, 2006, *Impacts of Increasing Ocean Acidification on Coral Reefs and Other Marine Calcifiers: A Guide for Future Research*, NSF, NOAA, and the U.S. Geological Survey, 88 pp., available at *http://www.isse.ucar.edu/florida/*.

Krabill, W., E. Hanna, P. Huybrechts, W. Abdalati, J. Cappelen, B. Csatho, E. Frederick, S. Manizade, C. Martin, J. Sonntag, R. Swift, R. Thomas, and J. Yungel, 2004, Greenland Ice Sheet: Increased coastal thinning, *Geophysical Research Letters*, **31**, L23302, doi:10.1029/2004GL021533.

Kulkarni, A.V., I.M. Bahuguna, B.P. Rathore, S.K. Singh, S.S. Randhawa, R.K. Sood, and S. Dhar, 2007, Glacial retreat in Himalaya using Indian remote sensing satellite data, *Current Science*, **92**, 69–74.

Kunkel, K.E., P.D. Bromirski, H.E. Brooks, T. Cavazos, A.V. Douglas, D.R. Easterling, K.A. Emanuel, P.Y. Groisman, G.J. Holland, T.R. Knutson, J.P. Kossin, P.D. Komar, D.H. Levinson, and R.L. Smith, 2008, Observed changes in weather and climate extremes, in *Weather and Climate Extremes in a Changing Climate. Regions of Focus: North America, Hawaii, Caribbean, and U.S. Pacific Islands*, Synthesis and Assessment Product 3.3, Climate Change Science Program and Subcommittee on Global Change Research, Washington, D.C., pp. 35–80.

Lagadec, P., 2004, Understanding the French 2003 heat wave experience: Beyond the heat, a multi-layered challenge, *Journal of Contingencies Crisis Management*, **12**, 160–169.

Lambert, F.H., A.R. Stine, N.Y. Krakauer, and J.C.H. Chiang, 2008, How much will precipitation increase with global warming? *EOS, Transactions of the American Geophysical Union*, **89**, 193–194.

Lemos, M.C., 2008, What influences innovation adoption by water managers? Climate information use in Brazil and the United States, *Journal of the American Water Resources Association*, **44**, 1388–1396.

Lemos, M., E. Boyd, E.L. Tompkins, H. Osbahr, and D. Liverman, 2007, Developing adaptation and adapting development, *Ecology and Society*, **12**, 26.

Lenton, T.M., H. Held, E. Kriegler, J.W. Hall, W. Lucht, S. Rahmstorf, and H.J. Schellnhuber, 2008, Tipping elements in the Earth's climate system, *Proceedings of the National Academy of Sciences*, **105**, 1786–1793.

Levine, J.B., and G.D. Salvucci, 1999, Equilibrium analysis of groundwater-vadose zone interactions and the resulting spatial distribution of hydrologic fluxes across a Canadian prairie, *Water Resources Research*, **35**, 1369–1383.

Levy, H., II, D.T. Shindell, A. Gilliland, M.D. Schwarzkopf, and L.W. Horowitz, eds., 2008, *Climate Projections Based on Emissions Scenarios for Long-Lived and Short-Lived Radiatively Active Gases and Aerosols*, Synthesis and Assessment Product 3.2, Climate Change Science Program and Subcommittee on Global Change Research, Washington, D.C., 100 pp.

Liu, S., Y. Ding, D. Shangguan, Y. Zhang, J. Li, H. Han, J. Wang, and C. Xie, 2006, Glacier retreat as a result of climate warming and increased precipitation in the Tarim river basin, northwest China, *Annals of Glaciology*, **43**, 91–96.

Malhi, Y., J.T. Roberts, R.A. Betts, T.J. Killeen, W. Li, and C.A. Nobre, 2008, Climate change, deforestation, and the fate of the Amazon, *Science*, **219**, 169–172.

Manuel, J, 2006, In Katrina's wake, *Environmental Health Perspectives*, **114**, A32–A39.

Manne, A.S., and R.G. Richels, 1991, Buying greenhouse insurance, *Energy Policy*, **19**, 543–562.

Mansur, E.T., R. Mendelsohn, and W. Morrison, 2008, Climate change adaptation: A study of fuel choice and consumption in the U.S. energy sector, *Journal of Environmental Economics and Management*, **55**, 175–193.

McCabe, G.J., M.P. Clark, and M.C. Serreze, 2001, Trends in Northern Hemisphere surface cyclone frequency and intensity, *Journal of Climate*, **14**, 2763–2768.

McFadden, D., 1984, Welfare analysis of incomplete adjustment to climatic change, in *Advances in Applied Microeconomics*, Vol. 3, V.K. Smith and A.D. Witte, eds., JAI Press, Greenwich, CT, pp. 133–149.

McMichael, A.J, D. Campbell-Lendrum, R.S. Kovats, S. Edwards, P. Wilkinson, N. Edmonds, N. Nicholls, S. Hales, F.C. Tanser, D. Le Sueur, M. Schlesinger, and N. Andronova, 2004, Global climate change, in *Comparative Quantification of Health Risks: Global and Regional Burden of Disease Due to Selected Major Risk Factors*, M. Ezzati, A. Lopez, A. Rodgers, and C. Murray, eds, World Health Organization, Geneva, pp. 1543–1649.

Meehl, G.A., T.F. Stocker, W.D. Collins, P. Friedlingstein, A.T. Gaye, J.M. Gregory, A. Kitoh, R. Knutti, J.M. Murphy, A. Noda, S.C.B. Raper, I.G. Watterson, A.J. Weaver, and Z.-C. Zhao, 2007, Global climate projections, in *Climate Change 2007: The Physical Basis*, Contribution of Working Group I to the Fourth Assessment Report of the Intergovernmental Panel on Climate Change, S. Solomon, D. Qin, M. Manning, Z. Chen, M. Marquis, K.B. Averyt, M. Tignor, and H.L. Miller, eds., Cambridge University Press, New York, pp. 747–845.

Meier, M.F., M.B. Dyurgerov, U.K. Rick, S. O'Neel, W.T. Pfeffer, R.S. Anderson, S.P. Anderson, and A.F. Glazovsky, 2007, Glaciers dominate eustatic sea-level rise in the 21st century, *Science*, **317**, 1064–1067.

Meinke, H., and R.C. Stone, 2005, Seasonal and interannual climate forecasting: The new tool for increasing preparedness to climate variability and change in agricultural planning operations, *Climate Change*, **70**, 221–253.

Mendelsohn, R., and J. Neumann, eds., 1999, *The Impact of Climate Change on the U.S. Economy*, Cambridge University Press, New York, 344 pp.

Mendelsohn, R., D. Shaw, and W. Nordhaus, 1994, The impact of global warming on agriculture: A Ricardian analysis, *The American Economic Review*, **84**, 753–771.

Mendelsohn, R., A. Basist, P. Kurukulasuriya, and A. Dinar, 2007, Climate and rural income, *Climatic Change*, **81**, 101–118.

Metcalf, G.E., 1999, A distributional analysis of green tax reforms, *National Tax Journal*, **52**, 665–681.

Miles, E.L., A.K. Snover, L.C. Whitely Binder, E.S. Sarachik, P.W. Mote, and N. Mantua, 2006, An approach to designing a national climate service, *Proceedings of the National Academy of Sciences*, **103**, 19,616–19,623.

Millennium Ecosystem Assessment, 2005, *Ecosystems and Human Well-Being: Synthesis*, Island Press, Washington, D.C., 160 pp.

Morgan, M.G., R. Cantor, W.C. Clark, A. Fisher, H.D. Jacoby, A.C. Janetos, A.P. Kinzig, J. Melillo, R.B. Street, and T.J. Wilbanks, 2005, Learning from the U.S. National Assessment of Climate Change Impacts, *Environmental Science and Technology*, **39**, 9023–9032.

Morgenstern, R., and W. Pizer, 2007, *Reality Check: The Nature and Performance of Voluntary Environmental Programs in the United States, Europe, and Japan*, Resources for the Future Press, Washington, D.C., 189 pp.

Mount, J., and R. Twiss, 2005, Subsidence, sea level rise, and seismicity in the Sacramento-San Joaquin Delta, *San Francisco Estuary Watershed Science*, **3**, available at *http://repositories.cdlib.org/jmie/sfews/vol3/iss1/art5/*.

Mukhala, E., and A. Chavula, 2007, Challenges to coping strategies with agrometeorological risks and uncertainties in Africa, in *Managing Weather and Climate Risks in Agriculture*, M.V.K. Sivakumar and R.P. Motha, eds., Springer, New York, pp. 39–49.

Murnane, R.J., 2004, Climate research and reinsurance, *Bulletin of the American Meteorological Society*, **85**, 697–707.

Najjar, R.G., H.A. Walker, P.J. Anderson, E.J. Barron, R.J. Bord, J.R. Gibson, V.S. Kennedy, C.G. Knight, J.P. Megonigal, R.E. O'Connor, C.D. Polsky, N.P. Psuty, B.A. Richards, L.C. Sorenson, E.M. Steele, and R.S. Swanson, 2000, The potential impacts of climate change on the mid-Atlantic coastal region, *Climate Research*, **14**, 219–233.

NAST (National Assessment Synthesis Team), 2001, *Climate Change Impacts on the United States: The Potential Consequences of Climate Variability and Change*, U.S. Global Change Research Program, Washington, D.C., 541 pp.

Negrón, J.F., B.J. Bentz, C.J. Fettig, N. Gillette, E.M. Hansen, J.L. Hayes, R.G. Kelsey, J.E. Lundquist, A.M. Lynch, R.A. Progar, and

S.J. Seybold, 2008, U.S. Forest Service bark beetle research in the western United States: Looking toward the future, *Journal of Forestry*, **106**, 325–331.

NOAA (National Oceanic and Atmospheric Administration), 2001, *The United States Detailed National Report on Systematic Observations for Climate: United States Global Climate Observing System (U.S.-GCOS) Program*, NESDIS, Rockville, MD, 138 pp.

Norby, R.J., E.H. DeLucia, B. Gielen, C. Calfapietra, C.P. Giardina, J.S. King, J. Ledford, H.R. McCarthy, D.J.P. Moore, R. Ceulemans, P. De Angelis, A.C. Finzi, D.F. Karnosky, M.E. Kubiske, M. Lukac, K.S. Pregitzer, G.E. Scarascia-Mugnozza, W.H. Schlesinger, and R. Oren, 2005, Forest response to elevated CO_2 is conserved across a broad range of productivity, *Proceedings of the National Academy of Sciences*, **102**, 18,052–18,056.

Nordhaus, W.D., 1977, Economic growth and climate: The carbon dioxide problem, *American Economic Review*, **67**, 341–346.

Nordhaus, W.D., 1991, To slow or not to slow: The economics of the greenhouse effect, *Economic Journal*, **101**, 920–937.

Nordhaus, W.D., 1994, *Managing the Global Commons: The Economics of Climate Change*, MIT Press, Cambridge, MA., 223 pp.

Nordhaus, W.D., 2006, The Economics of Hurricanes in the United States, NBER Working Paper 12813, National Bureau of Economic Research, Cambridge, MA, 46 pp.

Nordhaus, W., 2008, *A Question of Balance: Weighing the Options on Global Warming Policy*, Yale University Press, New Haven, CT, 256 pp.

NRC (National Research Council), 1992, *Global Environmental Change: Understanding the Human Dimensions,* P.C. Stern, O.R. Young, and D. Druckman, eds., National Academy Press, Washington, D.C., 320 pp.

NRC, 1997, *Environmentally Significant Consumption: Research Directions*, P.C. Stern, T. Dietz, V.W. Ruttan, R.H. Socolow, and J.L. Sweeney, eds., National Academy Press, Washington, D.C., 152 pp.

NRC, 1998, *People and Pixels: Linking Remote Sensing and Social Science*, National Academy Press, Washington, D.C., 256 pp.

NRC, 1999a, *Adequacy of Climate Observing Systems*, National Academy Press, Washington D.C., 51 pp.

NRC, 1999b, *Global Environmental Change: Research Pathways for the Next Decade*, National Academy Press, Washington, D.C., 616 pp.

NRC, 1999c, *Human Dimensions of Global Environmental Change: Research Pathways for the Next Decade*, National Academy Press, Washington, D.C., 100 pp.

NRC, 2000, *Issues in the Integration of Research and Operational Satellite Systems for Climate Research. Part I. Science and Design,* National Academy Press, Washington, D.C., 152 pp.

NRC, 2001, *A Climate Services Vision: First Steps Towards the Future,* National Academy Press, Washington D.C., 84 pp.

NRC, 2003, *Fair Weather: Effective Partnerships in Weather and Climate Services,* National Academies Press, Washington, D.C., 220 pp.

NRC, 2004a, *Climate Data Records from Environmental Satellites: Interim Report,* National Academies Press, Washington, D.C., 150 pp.

NRC, 2004b, *Groundwater Fluxes Across Interfaces,* National Academies Press, Washington, D.C., 100 pp.

NRC, 2004c, *Implementing Climate and Global Change Research: A Review of the Final U.S. Climate Change Science Program Strategic Plan,* National Academies Press, Washington, D.C., 96 pp.

NRC, 2005a, *Decision Making for the Environment: Social and Behavioral Science Research Priorities,* G.D. Brewer and P.C. Stern, eds., National Academies Press, Washington, D.C., 296 pp.

NRC, 2005b, *Thinking Strategically: The Appropriate Use of Performance Measures for the Climate Change Science Program,* National Academies Press, Washington, D.C., 150 pp.

NRC, 2007a, *Analysis of Global Change Assessments: Lessons Learned,* National Academies Press, Washington, D.C., 196 pp.

NRC, 2007b, *Earth Science and Applications from Space: National Imperatives for the Next Decade and Beyond,* National Academies Press, Washington, D.C., 456 pp.

NRC, 2007c, *Evaluating Progress of the U.S. Climate Change Science Program: Methods and Preliminary Results,* National Academies Press, Washington, D.C., 170 pp.

NRC, 2008a, *Earth Observations from Space: The First 50 Years of Scientific Achievements,* National Academies Press, Washington, D.C., 144 pp.

NRC, 2008b, *Ensuring the Climate Record from the NPOESS and GOES-R Spacecraft: Elements of a Strategy to Recover Measurement Capabilities Lost in Program Restructuring,* National Academies Press, Washington, D.C., 180 pp.

NRC, 2008c, *Review of CCSP Draft Synthesis and Assessment Product 5.3: Decision-Support Experiments and Evaluations Using Seasonal to Interannual Forecasts and Observational Data,* National Academies Press, Washington D.C., 56 pp.

NRC, 2008d, *Water Implications of Biofuels Production in the United States,* National Academies Press, Washington D.C., 88 pp.

NRC, 2009, *Informing Decisions in a Changing Climate,* National Academies Press, Washington D.C., in press.

Olsson, P., C. Folke, and F. Berkes, 2004, Adaptive comanagement for building resilience in social-ecological systems, *Environmental Management*, **34**, 75–90.

ORION Executive Steering Committee, 2005, *Ocean Observatories Initiative Science Plan*, Washington, D.C., 102 pp., available at *http://www.geo-prose.com/projects/ooi_sp.html.*

Orr, J.C., V.J. Fabry, O. Aumont, L. Bopp, S.C. Doney, R.A. Feely, A. Gnanadesikan, N. Gruber, A. Ishida, F. Joos, R.M. Key, K. Lindsay, E. Maier-Reimer, R. Matear, P. Monfray, A. Mouchet, R.G. Najjar, G.-K. Plattner, K.B. Rodgers, C.L. Sabine, J.L. Sarmiento, R. Schlitzer, R.D. Slater, I.J. Totterdell, M.-F. Weirig, Y. Yamanaka, and A. Yool, 2005, Anthropogenic ocean acidification over the twenty-first century and its impact on calcifying organisms, *Nature*, **437**, 681–686.

Pahl-Wostl, C., 2007, Transitions toward adaptive management of water facing climate and global change, *Water Resources Management*, **21**, 49–62.

Pandolfi, J.M., R.H. Bradbury, E. Sala, T.P. Hughes, K.A. Bjorndal, R.G. Cooke, D. McArdle, L. McClenachan, M.J.H. Newman, G. Paredes, R.R. Warner, and J.B.C. Jackson, 2003, Global trajectories of the long-term decline of coral reef ecosystems, *Science*, **301**, 955–958.

Parry, M.L., C. Rosenzweig, and M. Livermore, 2005, Climate change, global food supply and risk of hunger, *Philosophical Transactions of the Royal Society*, **360**, 2125–2138.

Patz, J.A., H.K. Gibbs, J.A. Foley, J.V. Rogers, and K.R. Smith, 2007, Climate change and global health: Quantifying a growing ethical crisis, *EcoHealth*, **4**, 397–405.

Pfeffer, W.T., J.T. Harper, and S. O'Neel, 2008, Kinematic constraints on glacier contributions to 21st-century sea-level rise, *Science*, **321**, 1340–1343.

Pielke, R.A., Jr., 2007, Future economic damage from tropical cyclones: Sensitivities to societal and climate changes, *Philosophical Transactions of the Royal Society* A, **365**, 2717–2729.

Pielke, R.A., Jr., J. Gratz, C.W. Landsea, D. Collins, M. Saunders, and R. Musulin, 2008, Normalized hurricane damages in the United States: 1900-2005, *Natural Hazards Review*, **9**, 29–42.

Pitman, G.K., 2002, *Bridging Troubled Waters: Assessing the World Bank Water Resources Strategy*, World Bank, Washington, D.C., 116 pp.

Potter, J.R., V.R. Burkett, and M.J. Savonis, 2008, Executive summary, in *Impacts of Climate Change and Variability on Transportation Systems and Infrastructure: Gulf Coast Study, Phase I*, M.J. Savonis, V.R. Burkett, and J.R. Potter, eds., Synthesis and Assessment Product 4.7, Climate Change Science Program and Subcommittee on Global Change Research, Washington, D.C., pp. ES1–ES10.

Rabalais, N.N., R.E. Turner, and W.J. Wiseman, 2002, Gulf of Mexico hypoxia, aka "The dead zone," *Annual Review of Ecology and Systematics*, **33**, 235–263.

Rabe, B., 2004, *Statehouse and Greenhouse: The Evolving Politics of American Climate Change Policy*, Brookings Institution Press, Washington, D.C., 236 pp.

Rahmstorf, S., 2007, A semi-empirical approach to projecting future sea level rise, *Science*, **315**, 368–370.

Rahmstorf, S., A. Cazenave, J.A. Church, J.E. Hansen, R.F. Keeling, D.E. Parker, and R.C.J. Somerville, 2007, Recent climate observations compared to projections, *Science*, **316**, 709.

Ramanathan, V., and Y. Feng, 2008, On avoiding dangerous anthropogenic interference with the climate system: Formidable challenges ahead, *Proceedings of the National Academy of Sciences*, **105**, 14,245–14,250.

Ramankutty, N., J.A. Foley, J. Norman, and K. McSweeney, 2002, The global distribution of cultivable lands: Current patterns and sensitivity to possible climate change, *Global Ecology and Biogeography*, **11**, 377–392.

Reilly, J., and D. Schimmelpfennig, 2000, Irreversibility, uncertainty and learning: Portraits of adaptation for long-term climate change, *Climatic Change*, **45**, 253–278.

Reilly, J., F. Tubiello, B. McCarl, and J. Melillo, 2000, Climate change and agriculture in the United States, in *Climate Change Impacts on the United States: The Potential Consequences of Climate Variability and Change*, U.S. Global Change Research Program, Washington, D.C., pp. 379–403.

Rignot, E., and P. Kanagaratnam, 2006, Changes in the velocity structure of the Greenland Ice Sheet, *Science*, **311**, 986–990.

Rignot, E., J.L. Bamber, M.R. van den Broeke, C. Davis, Y. Li, W.J. van de Berg, and E. Van Meijgaard, 2008, Recent Antarctic ice mass loss from radar altimetry and regional climate modeling, *Nature Geoscience*, **1**, 106–110.

Rosenfeld, D., U. Lohmann, G.B. Raga, C.D. Dowd, M. Kulmala, S. Fuzzi, A. Reissell, and M.O. Andreae, 2008, Flood or drought: How do aerosols affect precipitation? *Science*, **321**, 1309–1313.

Santer, B.D., C. Mears, F.J. Wentz, K.E. Taylor, P.J. Gleckler, T.M.L. Wigley, T.P. Barnett, J.S. Boyle, W. Brüggermann, N.P. Gillett, S.A. Klein, G.A. Meehl, T. Nozawa, D.W. Pierce, P.A. Stott, W.M. Washington, and M.F. Wehner, 2007, Identification of human-induced changes in atmospheric moisture content, *Proceedings of the National Academy of Sciences*, **104**, 15,248–15,253.

Savonis, M.J., V.R. Burkett, and J.R. Potter, eds., 2008, *Impacts of Climate Change and Variability on Transportation Systems and Infrastructure: Gulf Coast Study, Phase I*, Synthesis and Assessment Product 4.7, Climate Change Science Program and Subcommittee on Global Change Research, Washington, D.C., 445 pp.

Scambos, T.A., J.A. Bohlander, C.A. Shuman, and P. Skvarca, 2004, Glacier acceleration and thinning after ice shelf collapse in the Larsen B embayment, Antarctica, *Geophysical Research Letters*, **31**, L18402, doi:10.1029/2004GL020670.

Schlenker, W., W.M. Hanemann, and A.C. Fisher, 2005, Will U.S. agriculture really benefit from global warming? Accounting for irrigation in the hedonic approach, *American Economic Review*, **95**, 395–406.

Schliebe, S., T. Evans, K. Johnson, M. Roy, S. Miller, C. Hamilton, R. Meehan, and S. Jahrsdoerfer, 2006, *Range-Wide Status Review of the Polar Bear (Ursus maritimus)*, 262 pp., available at *http://alaska.fws.gov/fisheries/mmm/polarbear/pdf/Polar_Bear_%20Status_Assessment.pdf*.

Schmittner, A.O., D. Archer, H.D. Matthews, and E.D. Galbraith, 2008, Future changes in climate, ocean circulation, ecosystems, and biogeochemical cycling simulated for a business-as-usual CO_2 emission scenario until year 4000 AD, *Global Biogeochemical Cycles*, **22**, GB1013, doi:10.1029/2007GB002953.

Schneider, S.H., and M.D. Mastrandrea, 2005, Probabilistic assessment of "dangerous" climate change and emissions pathways, *Proceedings of the National Academy of Sciences*, **102**, 15,728–15,735.

Shea, E.L., G. Dolcemascolo, M.P. Hamnett, C. Anderson, N. Lewis, J. Loschnigg, T. Barnston, G. Meehl, C. Guard, and L. He, 2001, *Pacific Island Regional Assessment of the Consequences of Climate Change and Variability*, East-West Center, Honolulu, available at *http://www2.eastwestcenter.org/climate/assessment/report.htm*.

Soniat, T.M., E.E. Hofmann, J.M. Klinck, and E.N. Powell, 2008, Differential modulation of eastern oyster (*Crassostrea virginica*) disease parasites by the El-Niño-Southern Oscillation and the North Atlantic Oscillation, *International Journal of Earth Sciences*, **98**, 99–114.

Stern, N., 2006, *The Economics of Climate Change*, Cambridge University Press, Cambridge, 579 pp.

Thomas, R., E. Rignot, G. Casassa, P. Kanagaratnam, C. Acuña, T. Akins, H. Brecher, E. Frederick, P. Gogineni, W. Krabill, S. Manizade, H. Ramamoorthy, A. Rivera, R. Russell, J. Sonntag, R. Swift, J. Yungel, and J. Zwally, 2004, Accelerated sea-level rise from West Antarctica, *Science*, **306**, 255–258.

Thomas, R., E. Frederick, W. Krabill, S. Manizade, and C. Martin, 2006,

Progressive increase in ice loss from Greenland, *Geophysical Research Letters*, **33**, L10503, doi:10.1029/2006GL026075.

Titus, J.G., 1990, Greenhouse effect, sea level rise, and barrier islands: Case study of Long Beach Island, New Jersey, *Coastal Management*, **18**, 65–90.

Tompkins, E.L., M.C. Lemos, and E. Boyd, 2008, A less disastrous disaster: Managing response to climate-driven hazards in the Cayman Islands and NE Brazil, *Global Environmental Change*, **18**, 736–745.

Trenberth, K.E., P.D. Jones, P. Ambenje, R. Bojariu, D. Easterling, A. Klein Tank, D. Parker, F. Rahimzadeh, J.A. Renwick, M. Rusticucci, B. Soden, and P. Zhai, 2007, Observations: Surface and atmospheric climate change, in *Climate Change 2007. The Physical Science Basis*, Contribution of WG 1 to the Fourth Assessment Report of the Intergovernmental Panel on Climate Change, S. Solomon, D. Qin, M. Manning, Z. Chen, M.C. Marquis, K.B. Averyt, M. Tignor, and H.L. Miller, eds., Cambridge University Press, New York, pp. 235–336 plus online annex.

Trenberth, K.E., T. Koike, and K. Onogi, 2008, Progress and prospects in reanalysis for weather and climate, *EOS, Transactions of the American Geophysical Union*, **89**, doi:10.1029/2008EO260002.

Trostle, R., 2008, *Global Agricultural Supply and Demand: Factors Contributing to Recent Increase in Food Commodity Prices*, USDA Outlook Report WRS 0801, 30 pp., available at *http://www.ers.usda.gov/Publications/WRS0801/*.

Tubiello, N.F., and G. Fischer, 2007, Reducing climate change impacts on agriculture: Global and regional effects of mitigation, 2000–2080, *Technological Forecasting and Social Change*, **74**, 1030–1056.

Tubiello, N.F., J.F. Soussanas, and S.M. Howden, 2007, Crop and pasture response to climate change, *Proceedings of the National Academy of Sciences*, **105**, 19,686–19,690.

UNDP (United Nations Development Programme), 2006, *Beyond Scarcity: Power, Poverty and the Global Water Crisis*, Human Development Report 2006, Palgrave McMillan, New York, 422 pp.

Uppala, S.M., P.W. Kallberg, A.J. Simmons, U. Andrae, V.D. Bechtold, M. Fiorino, J.K. Gibson, J. Haseler, A. Hernandez, G.A. Kelly, X. Li, K. Onogi, S. Saarinen, N. Sokka, R.P. Allan, E. Andersson, K. Arpe, M.A. Balmaseda, A.C.M. Beljaars, L. Van De Berg, J. Bidlot, N. Bormann, S. Caires, F. Chevallier, A. Dethof, M. Dragosavac, M. Fisher, M. Fuentes, S. Hagemann, E. Holm, B.J. Hoskins, L. Isaksen, P.A.E.M. Janssen, R. Jenne, A.P. McNally, J.F. Mahfouf, J.J. Morcrette, N.A. Rayner, R.W. Saunders, P. Simon, A. Sterl, K.E. Trenberth, A. Untch, D. Vasiljevic, P. Viterbo, and J. Woollen, 2005,

The ERA-40 reanalysis, *Quarterly Journal of the Royal Meteorological Society*, **131,** 2961–3012.

USDA (U.S. Department of Agriculture), 2007, *Ethanol Expansion in the United States: How Will the Agricultural Sector Adjust?* USDA FDS 07D01, Washington, D.C., 18 pp.

Vandentorren, S., and P. Empereur-Bissonnet, 2005, Health impacts of the 2003 heat wave in France, in *Extreme Weather Events and Public Health Responses*, W. Kirch, B. Nenne, and R. Bertollini, eds., Springer, New York, pp. 81–88.

Wang, X.L., V.R. Swail, and F.W. Zwiers, 2006, Climatology and changes of extratropical storm tracks and cyclone activity: Comparison of ERA-40 with NCEP/NCAR reanalysis for 1958-2001, *Journal of Climate*, **19**, 3145–3166.

Waugh, W.L., Jr., ed., 2006, *Shelter from the Storm: Repairing the National Emergency Management System After Katrina*, Annals of the American Academy of Political and Social Science, **604**, 322 pp.

Weitzman, M., 2009, On modeling and interpreting the economics of catastrophic climate change, *Review of Economics and Statistics*, **91**, 1–19.

Werick, W.J., and R.N. Palmer, 2008, It's time for standards of practice in water resources planning, *Journal of Water Resources Planning and Management*, **Jan/Feb**, 1–2.

West, S.E., and R.C. Williams III, 2004, Estimates from a consumer demand system: Implications for the incidence of environmental taxes, *Journal of Environmental Economics and Management*, **47**, 535–558.

Weyant, J.P., F.C. de la Chesnaye, and G.J. Blanford, 2006, Overview of EMF-21: Multigas mitigation and climate policy, *Energy Journal*, **27**, 1–32.

Wier, M., K. Birr-Pedersen, H.K. Jacobsen, and J. Klok, 2005, Are CO_2 taxes regressive? Evidence from the Danish experience, *Ecological Economics*, **52**, 239–251.

Wigley, T.M.L., R. Richels, and J.A. Edmonds, 1996, Economic and environmental choices in the stabilization of atmospheric CO_2 concentrations, *Nature*, **379**, 240–243.

Wijkman, A., and L. Timberlake, 1984, *Natural Disasters: Acts of God or Acts of Man?* Earthscan Ltd., Washington, D.C., 150 pp.

Wilbanks, T.J., P. Romero Lankao, M. Bao, F. Berkhout, S. Cairncross, J.-P. Ceron, M. Kapshe, R. Muir-Wood, and R. Zapata-Marti, 2007, Industry, settlement and society, in *Climate Change 2007: Impacts, Adaptation and Vulnerability*, Contribution of Working Group II to the Fourth Assessment Report of the Intergovernmental Panel on Climate Change, M.L. Parry, O.F. Canziani, J.P. Palutikof, P.J. van der Linden, and C.E. Hanson, eds., Cambridge University Press, Cambridge, pp. 357–390.

Wisner, B., P. Blaikie, T. Cannon, and I. Davis, 2004, *At Risk. Natural Hazards, People's Vulnerability and Disasters*, Routledge, London, 471 pp.

Yohe, G., and J. Neumann, 1997, Planning for sea level rise and shore protection under climate uncertainty, *Climatic Change*, **37**, 111–140.

Zellner, M.L., S.E. Page, W. Rand , D.G. Brown, D.T. Robinson, J. Nassauer, and B. Low, 2008, The emergence of zoning policy games in exurban jurisdictions: Informing collective action theory, *Land Use Policy*, **26**, 356–367.

Zeng, N., 2003, Drought in the Sahel, *Science*, **302**, 999–1000.

Zhao, T., D.G. Brown, and K.M. Bergen, 2007, Increasing gross primary production (GPP) in the urbanizing landscape of southeastern Michigan, *Photogrammetric Engineering and Remote Sensing*, **73**, 1159–1168.

Appendixes

Appendix A

Examples of Bills with a Significant Climate Change Component Considered in the 110th Congress

Bill	Comments
HR 6—Energy Independence and Security Act of 2007 (P.L. 110-40)	Omnibus energy act (enacted). Among other things, directs DOE to carry out research and development relating to alternative energy sources, including biofuels and geothermal. Directs DOE to carry out carbon capture and sequestration research and development and demonstration programs, authorizing $1.4B in appropriations for these actions.
S 280—Climate Stewardship and Innovation Act of 2007	Would establish program for market-driven reduction of greenhouse gases via emissions trading program, would require EPA to establish greenhouse gas database. Also would authorize $60M in appropriations for NOAA abrupt climate change research. (HR 4266 was the House version.)

Bill	Comments
HR 906—Global Change Research and Data Management Act of 2007	Would reauthorize the U.S. Global Change Research Program to focus on decision support needs, and would require preparation of national vulnerability assessments on a 5-year cycle. Would also require preparation of a report to Congress on coordinating federal climate data management and archiving.
S 1018/HR 1961— Global Climate Change Security Oversight Act	Would require DOD, DOS, and ODNI to take specified actions related to estimating effects of climate change on national security, would authorize DOD research on climate change.
S 1581/HR 4174— Federal Ocean Acidification Research and Monitoring Act of 2007	Would direct DOC to establish a NOAA ocean acidification research program, authorizing $70M in appropriations. Another Senate version is S 2211.
S 1874— Containing and Managing Climate Costs Efficiently Act	Would establish a carbon market efficiency board to analyze carbon market information and to intervene in such a market.
S 2156—Science and Engineering to Comprehensively Understand and Responsibly Enhance Water Act	Would direct DOI to establish a USBR climate change adaptation program, including authorizing the funding of water resources research, and would direct DOE to assess risk of climate change to hydropower generation.

Bill	Comments
S 2191—America's Climate Security Act of 2007 ("Lieberman bill")	Would require EPA to establish a greenhouse gas registry and trading system, would also establish a carbon market efficiency board.
S 2204—Global Warming Wildlife Survival Act	Would require DOI and DOC to establish national strategies for maintaining wildlife populations/marine ecosystems in view of climate change effects, and would require USGS to examine effects of climate change on listed species.
HR 2338—Global Warming Wildlife Survival Act	Would require DOI to promulgate a national strategy for mitigating climate change impacts on wildlife. Would require establishment of a science program within USGS to conduct research on impacts of warming on wildlife and habitat, and to provide science support to federal resource management agencies on such topics.
HR 2342—National Integrated Coastal and Ocean Observation Act of 2008	Would direct the president to establish a coastal and ocean observation system to, among other things, improve the ability to track and predict climate variability and change.
S 2355—Climate Change Adaptation Act	Among other things, would establish a national climate service within NOAA, would require DOC to prepare regional assessments on coastal and ocean vulnerability to climate change, and would require DOC to provide grants to coastal states for developing coastal and ocean adaptation plans.
S 2970/HR 6297—Climate Change Drinking Water Adaptation Research Act	Would require EPA, in cooperation with DOC, DOE, and DOI, to fund an applied research program on drinking water utility adaptation to climate change. Would authorize $275M in appropriations.

Bill	Comments
HR 5453—Coastal State Climate Change Planning Act of 2008	Would establish a DOC coastal climate change adaptation planning and response program, and would authorize grants to coastal states for implementation.
S 2307—Global Change Research Improvement Act of 2007	Would amend the Global Change Research Act to establish a National Climate Service within NOAA.

NOTES: DOC = Department of Commerce; DOD = Department of Defense; DOE = Department of Energy; DOI = Department of Interior; DOS = Department of State; EPA = Environmental Protection Agency; NOAA = National Oceanic and Atmospheric Administration; ODNI = Office of the Director of National Intelligence; USBR = U.S. Bureau of Reclamation; USGS = U.S. Geological Survey.

Appendix B

U.S. Climate Change Science Program

The U.S. Climate Change Science Program (CCSP) integrates the U.S. Global Change Research Program (USGCRP) and the Climate Change Research Initiative (CCRI). The USGCRP, the first federally coordinated program supporting climate change research, began as a presidential initiative in 1988 and received congressional support in 1990 under the Global Change Research Act. The act called for the development of a research program "to understand, assess, predict, and respond to human-induced and natural processes of global change," and it guided federally supported global change research for the next decade. In 2001, President Bush launched the CCRI to investigate uncertainties and set new research priorities in climate change science. The CCRI also gave priority to research that could yield results within a few years, either by improving decision-making capabilities or by contributing to improved public understanding. The two programs were merged into the CCSP the following year and given an ambitious guiding vision: *a nation and the global community empowered with the science based knowledge to manage the risks and opportunities of change in the climate and related environmental systems.*

The CCSP is guided by five overarching goals and organized into seven research elements and six crosscutting issues (CCSP, 2003):

Overarching goals

1. Improve knowledge of the Earth's past and present climate and environment, including its natural variability, and improve understanding of the causes of observed variability and change

2. Improve quantification of the forces bringing about changes in the Earth's climate and related systems

3. Reduce uncertainty in projections of how the Earth's climate and related systems may change in the future

4. Understand the sensitivity and adaptability of different natural and managed ecosystems and human systems to climate and related global changes

5. Explore the uses and identify the limits of evolving knowledge to manage risks and opportunities related to climate variability and change

Research elements: Atmospheric composition, climate variability and change, water cycle, land-use/land-cover change, carbon cycle, ecosystems, and human contributions and responses to environmental change

Crosscutting issues: Decision support resources development, communications, modeling strategy, observing and monitoring the climate system, data management, and international cooperation

The CCSP research elements are consistent with but broader than those of the predecessor U.S. Global Change Research Program. A time line of how the research focus has evolved is given in Table B.1.

TABLE B.1 Evolution of Research Elements, USGCRP and CCSP

USGCRP			CCSP			
1989-1996	1997-1999	2000	2001	2002	2003	2004-2008
Climate and hydrologic systems	Climate change over decades to centuries	Climate system	Climate system	Climate system	Climate system	Climate variability and change
	Seasonal to interannual climate variability					
		Global water cycle	Global water cycle	Global water cycle	Global water cycle	Global water cycle
Biogeochemical dynamics			Global carbon cycle	Global carbon cycle	Global carbon cycle	Global carbon cycle
Ecological systems and dynamics	Changes in land cover and in terrestrial and marine ecosystems		Biology and biogeochemistry of ecosystems	Biology and biogeochemistry of ecosystems	Biology and biogeochemistry of ecosystems	Ecosystems

TABLE B.1 Continued

USGCRP				CCSP		
1989-1996	1997-1999	2000	2001	2002	2003	2004-2008
					Land use/land cover change	Land-use/land-cover change
Human interactions	Human contributions and responses to global change	Human dimensions of global change	Human dimensions of global change	Human dimensions of global change	Human dimensions of global change	Human contributions and responses to environmental change
Solar influences	Changes in ozone, ultraviolet radiation, and atmospheric chemistry	Composition and chemistry of the atmosphere	Composition and chemistry of the atmosphere	Composition and chemistry of the atmosphere	Composition and chemistry of the atmosphere	Atmospheric composition
Earth system history		Paleoenvironment and paleoclimate	Paleoenvironment and paleoclimate			
Solid Earth processes						

SOURCES: CCSP (2003, *Strategic Plan*; 1989–2008, *Our Changing Planet*).

The CCSP is managed by a director with the help of a program office and interagency committees that plan future research and crosscutting activities (e.g., decision support, communications). Funding is controlled and managed by the individual participating agencies and has been declining since 1996, mostly because of decreases in NASA's investment in climate observations (Figure 1.2). Participating agencies include the Agency for International Development, Department of Agriculture, Department of Defense, Department of Energy, Department of State, Department of Transportation, Environmental Protection Agency, National Aeronautics and Space Administration, National Institutes of Health, National Oceanic and Atmospheric Administration, National Science Foundation, Smithsonian Institution, and the U.S. Geological Survey.

REFERENCES

CCSP (Climate Change Science Program), 1989–2008, *Our Changing Planet: The U.S. Climate Change Science Program*, Climate Change Science Program and Subcommittee on Global Change Research, Washington, D.C., 17 volumes.

CCSP, 2003, *Strategic Plan for the U.S. Climate Change Science Program*, Climate Change Science Program and Subcommittee on Global Change Research, Washington, D.C., 202 pp.

Appendix C

Process for Identifying Priority Areas

Considerable thinking has been done on how to set science priorities in the federal government (see summary in NRC, 2006), going back before Alvin Weinberg's two benchmark articles on criteria for scientific choice (Weinberg, 1963, 1964). This appendix describes the process the committee used to identify priorities for a future climate research program. The committee's approach was informed by the decision science literature and modeled after the method used to identify risk reduction options for the Environmental Protection Agency (EPA, 1990). The steps the committee followed were:

1. Specify the goals that the climate research and applications are intended to achieve
2. Identify the major priority areas
3. Develop criteria for ranking the priority areas
4. Convene two stakeholder workshops—the first on applications and the second on science—to revise and rank the priority areas
5. Choose a final list of priorities based on workshop and other input and connections with climate change issues of importance to society

The first four steps were done in an iterative consensus process in which a strawman list was vetted and modified by outside experts in several rounds of discussion at committee meetings and workshops. Although such methods have well-known shortcomings (e.g., validity, reliability, problems concerning the consensus among the experts), they have proven useful when it is not possible to obtain objective data (Finkel and Golding, 1994; Davies, 1996). The last step was carried out by the committee, which is responsible for the priorities presented in this report.

IDENTIFYING SCIENCE AND APPLICATIONS PRIORITY AREAS

The overarching goals of the Climate Change Science Program (CCSP; Appendix B) provided the context for identifying both science and applications priority areas (Step 1; Table C.1). Strawman priority areas (Step 2) were gleaned from workshops and more than 100 published reports and articles to give them a level of community review and acceptance. Among the most important sources were the gaps and weaknesses identified in *Evaluating Progress of the U.S. Climate Change Science Program: Methods and Preliminary Results* (NRC, 2007) and discussion papers prepared by the National Academies Committee on the Human Dimensions of Global Change (CHDGC) and Climate Research Committee (CRC; see Appendixes D and E). The CHDGC and CRC narrowed down dozens of candidate priorities using criteria similar to those developed by the committee for Step 3 and feedback from the committee.

TABLE C.1 Primary Sources of Input for Prioritization Process

Topic	Step 1 Goals	Step 2 Priority Areas	Step 3 Criteria	Step 4 Workshop
Applications	CCSP overarching goals	NRC (2007) and CHDGC and CRC discussion papers	Workshop participants	October 15–17, 2007; 104 experts: • 37% academia • 11% industry • 44% government • 8% NGO
Science	CCSP overarching goals	CHDGC and CRC discussion papers	22 NRC reports on setting science priorities	March 19–20, 2008; 78 experts: • 69% academia • 5% industry • 17% government • 9% NGO

NOTE: CHDGC = Committee on the Human Dimensions of Global Change; CCSP = Climate Change Science Program; CRC = Climate Research Committee; NGO = nongovernmental organization; NRC = National Research Council.

Criteria for ranking science priority areas (Step 3) were compiled from 22 National Research Council reports and the most important ones were chosen and revised at the science workshop. The final criteria were:

- Scientific merit (e.g., generates new knowledge or fills critical gaps)
- Readiness (scientific, technical, programmatic, community capacity)
- Impacts (e.g., breadth of beneficiaries; potential for informing decisions, improving public understanding, or reducing risk)
- Cost

The criteria for ranking applications priority areas were identified at the applications workshop. No attempt was made to develop a common set of criteria.

Step 4 (refining and ranking the priority areas) was accomplished at the workshops. In each workshop, plenary discussions alternated with working group sessions, allowing multiple opportunities for input and iteration with diverse groups over the course of 2 or 3 days. Both workshops included a mix of experts from academia, industry, government, and nongovernmental organizations, but the applications workshop also included congressional staff and the media and had a much higher proportion of social scientists and managers from industry and federal, state, and local government (Table C.1). Natural and social scientists dominated the science workshop. Priority areas that emerged from the two workshops are listed, in random order, in Boxes C.1, C.2, and C.3.

The workshops ranked priority areas within applications, natural science, and human dimensions. A final list of priorities from among these different areas (Step 5) and other sources was selected by the committee and appears in Chapter 3.

**BOX C.1 Application Priority Areas From
the October 2007 Workshop**

- Climate prediction, rather than Intergovernmental Panel on Climate Change-style scenarios
- Increased model resolution to improve predictive capabilities, especially at decadal (e.g., over the next 10 to 20 years) and regional/local scales
- Projection of additional variables required by specific user groups (e.g., temperature, humidity, wind for fire hazard; precipitation duration, intensity, and phase for water managers)
- Accounting for the tails of the probability distribution of future climate changes for risk analysis, adaptation, and cost estimates of climate changes (scientists tend to work with the means)
- Regional climate change impact analysis (i.e., the locations of various changes in different regions and the time frames over which the changes will occur)
- National integrated assessment of climate change, focused on impacts, adaptation (natural, land use, and social systems), and multiple stressors (e.g., climate in a socioeconomic context)
- Operational climate services to create climate assessments and predictions and to provide information tailored to different user communities
- Establishment of a data, tool, and scenario library or clearinghouse for climate data and information users

---

**BOX C.2 Natural Science Priority Areas From
the March 2008 Workshop**

- Integrated Earth system analysis (analysis, prediction, and evaluation testing of models against data)
- Land-use change, including carbon cycle and land management in the context of mitigation and adaptation
- Ocean parameterizations (e.g., mixing processes, biological feedbacks, air–sea exchange)
- Modeling and observations of the tropics (convection, tropical storms, regional change)
- Impacts of increasing CO_2 levels on the oceans (including ocean acidification and marine ecosystems)
- Melting ice sheets, alpine glaciers, and sea level rise (including coastal impacts)
- Decadal prediction, with a focus on regional scales, including abrupt climate change
- Extreme events and hazards (especially hurricanes and drought)
- Land hydrological sensitivity to climate change (including drought and mountain glacier runoff/impact)
- End-to-end systems analysis/consequences of mitigation measures (including geoengineering and carbon sequestration)
- Modeling longer-timescale feedbacks
- Aerosols, clouds, precipitation, and atmospheric chemistry (connection to climate forcing and air quality)
- Operational attribution (not just global warming; connection to individual regional events)

**BOX C.3 Human Dimensions Priority Areas From
the March 2008 Workshop**

- Human drivers of change (e.g., migration and population growth, land use, lifestyle, household consumption) and their role in emissions generation, impact vulnerability, and adaptation
- Characterize adaptation, including autonomous (private) and public adaptation. This includes the role of social networks and institutions as well as the cost and speed of adaptation.
- Human health consequences of changes in the weather, ecosystems, and air pollution, including short- vs. long-run (adaptive) responses, extreme events vs. average conditions, and how these responses might be affected by migration, land-use change, and lifestyle
- Systems interactions and net effects of human behavior (mitigation and adaptation) on water, land, and energy use, carbon fluxes, ecosystems, coastal resources, and the built environment (methodological systems approach)
- Institutional and social constraints and opportunities for technological innovation, diffusion, and adoption in the context of mitigation (including geoengineering) and adaptation
- Characterizing human perceptions and valuations of impacts and risks of climate change, including variability, speed of change, and abrupt change
- Human and systems differential vulnerabilities to climate change including scenarios, mapping, and development of metrics of adaptive capacity
- Methods and processes to support effective climate decision making (including communication and education)
- Ethics and equity of climate change and responses

REFERENCES

Davies, J.C., ed., 1996, *Comparing Environmental Risks: Tools for Setting Government Priorities*, Resources for the Future, Washington, D.C., 157 pp.

EPA (Environmental Protection Agency), 1990, *Reducing Risk: Setting Priorities and Strategies for Environmental Protection*, Report by the Science Advisory Board, SAC-EC-90-021, Washington, D.C., 26 pp.

Finkel, A.M., and D. Golding, eds., 1994, *Worst Things First? The Debate over Risk-Based National Environmental Priorities*, Resources for the Future, Washington, D.C., 348 pp.

NRC (National Research Council), 2006, *A Strategy for Assessing Science: Behavioral and Social Research on Aging*, National Academies Press, Washington, D.C., 176 pp.

NRC, 2007, *Evaluating Progress of the U.S. Climate Change Science Program: Methods and Preliminary Results*, National Academies Press, Washington, D.C., 170 pp.

Weinberg, A., 1963, Criteria for scientific choice, *Minerva*, **1**, 159–171.

Weinberg, A., 1964, Criteria for scientific choice II: The two cultures, *Minerva*, **3**, 3–14.

Appendix D

Fundamental Research Priorities to Improve the Understanding of Human Dimensions of Climate Change

Paul C. Stern, National Research Council
Thomas J. Wilbanks, Oak Ridge National Laboratory

Note: The committee commissioned the following discussion paper from the staff and chair of the National Research Council Committee on the Human Dimensions of Global Change. Their views, as expressed below, may not always reflect the views of their committee, the Committee on Strategic Advice on the U.S. Climate Change Research Program, or vice versa.

INTRODUCTION

The Assignment

At the request of the U.S. Climate Change Science Program (CCSP), the National Research Council (NRC) has established a Committee on Strategic Advice on the U.S. Climate Change Science Program, charged with two tasks. Task 1 was to evaluate progress of the CCSP, and that report was completed in 2007 (NRC, 2007a). Task 2 is to provide advice to CCSP on future research priorities, and the key step in this process will be a national workshop on "discovery science" in March 2008.

One of the key findings of the Task 1 preliminary assessment was that "*our understanding of the impact of climate changes on human well-being and vulnerabilities ... is much less developed than our understanding of the natural climate system,*" a conclusion that echoed findings of the earlier NRC review of the CCSP Draft Strategic Plan (NRC, 2004). For the March workshop, the Committee on Strategic Advice on the U.S. Climate Change Science Program commissioned two discussion papers on research priorities for climate change science. At least partly reflecting the finding from its first report, one of the papers is focused on underlying research priorities for human systems science, including the social sciences. The other is an equivalent summary of priorities related to the natural sciences.

As initially articulated by Strategic Advice committee member Charles Kolstad, the assignment was to prepare a "paper on social science priorities" as an input to the workshop, identifying up to 10 top priorities and considering ways to increase the engagement of core disciplines as well as multidisciplinary researchers. Thus defined, the priorities were to be focused on relatively basic research rather than applied research. For the assignment, the Committee on Strategic Advice to the U.S. Climate Change Science Program enlisted the assistance of members of the NRC Committee on Human Dimensions of Global Change (CHDGC), who discussed the assignment in detail at the November 2007 CHDGC meeting. The result was a draft paper—an informal communication from the staff director of CHDGC.

The draft was discussed at the January meeting of the Strategic Advice committee, which asked that its scope be expanded to add an additional set of research priorities lying closer to the interests of mission agencies in the CCSP and comments on some implementation issues. This paper is the result.

The Terminology

Following the usual practice of CHDGC reports, this paper uses the terms *fundamental research* and *human dimensions* rather than *basic research* and *social science*:

Fundamental Research

In conventional usage, basic research is motivated by intellectual curiosity and undertaken for the "pure" pursuit of knowledge, not for social aims. Most of the basic research in the social and behavioral sciences is not motivated by climate concerns, and much of it has no obvious climate applications. Much the same may be true of basic research in chemistry or physics. A different kind of discovery science, equally concerned with advancing knowledge, derives its priorities from social needs and related programs (i.e., "purposive basic research") and has been termed "fundamental" research (Shapley and Roy, 1985). The two kinds of research are virtually identical in how they proceed; where they differ is how research questions are developed. We believe that research advice to CCSP is more appropriately considered the latter.

Human Dimensions Research

Throughout its 19-year history, in its attention to climate change as a special case of global environmental change, the CHDGC has been concerned with *human systems drivers of climate change, human systems impacts of climate change, and human systems responses to concerns about or observed effects of climate change*. These topics are grounded in the social, economic, and behavioral sciences but are not limited to these sciences. For example, driving forces include technologies, and so understanding them requires engineering expertise. Impacts include effects on human health, food, and energy systems, and understanding the processes producing such impacts requires knowledge and expertise beyond the social sciences alone. From the inception of the CHDGC (see NRC, 1992), its reports, which reflect the views of many human dimensions researchers, have identified research priorities for human dimensions of climate change and other kinds of global environmental change in terms of the *ends* of knowledge—what it is that requires understanding—rather than in terms of an arbitrarily constrained set of academic or disciplinary *means* for reaching the ends. In this paper, we have adopted an integrative approach rather than a disciplinary approach, as in past NRC reports.

Human Dimensions as a Distinct Interdisciplinary Field

Many scientists who conduct fundamental research on human–environment interactions conceive of the area as a distinct interdisciplinary field or even a distinct discipline. Various names have been proposed for this field, including human ecology, human–environment science (Stern, 1993), and more recently sustainability science (NRC, 1999d) and coupled human and natural systems (e.g., Liu et al., 2007a, b). The pathbreaking NRC report, *Our Common Journey* (NRC, 1999d), has led to such significant steps as the establishment of a new membership section in the National Academy of Sciences and the creation of a new section in *Proceedings of the National Academy of Sciences*. The editors of *PNAS* are actively promoting research in sustainability science and refer to it in the journal as a discipline. The National Science Foundation (NSF) has recently established a multidirectorate program on The Dynamics of Coupled Natural and Human Systems to support "quantitative, interdisciplinary analyses of relevant human and natural system processes and complex interactions among human and natural systems at diverse scales."[1]

A "Road Map" to This Paper

The paper is in four parts. It first considers the broader historical context within which priorities have been identified in the past and are identified here. Second, it identifies five substantive priorities for fundamental research and three crosscutting fundamental research issues, discusses criteria used to identify the priorities, and identifies the benefits that can result from the research. Third,

---

[1] *http://www.nsf.gov/pubs/2006/nsf06587/nsf06587.htm.*

it identifies climate change research priorities focused on human dimensions that are somewhat less fundamental or more action oriented, and shows some of the linkages between fundamental research and these priorities. Fourth, it identifies critical constraints on progress with these research topics, including but not limited to issues in relating to core disciplines, and offers some possible implementation strategies for overcoming these constraints.

THE CONTEXT FOR PRIORITY SETTING

There is a relatively rich history of efforts to set priorities for research on the human dimensions of climate change. These efforts provide a strong basis for identifying priorities, and this paper builds on that work. However, the history of responses to past priority-setting exercises shows that careful priority setting alone has made little difference, either in the behavior of agencies that might fund the recommended research or in attracting increased interest from several of the core disciplines in the behavioral and social sciences. These observations suggest that getting research supported and done requires more than identifying priorities. This issue is discussed below under "Critical Constraints."

History of Priority-Related Discussions

For almost two decades, committees and panels of the NRC have considered priorities for research on the human dimensions of global environmental change and/or global sustainability. Multiple major studies have helped to provide intellectual foundations for the field, and many others identify research priorities for all or part of the field. Each of these studies involved the participation of large numbers of professionals and stakeholders and produced reports that were extensively reviewed by peers. Many of them also included workshops to engage a larger community than the committee membership alone. The research recommendations from these studies provide a valuable base for this paper. An incomplete listing of these studies follows.

Publications primarily developing the intellectual basis for progress include:

- *The Drama of the Commons* (NRC, 2002c) summarized knowledge on major questions about the design and operation of institutions for managing common-pool resources band set research directions for the future.
- *New Tools for Environmental Protection: Education, Information, and Voluntary Measures* (NRC, 2002b) summarized available knowledge and examined the potential for these measures as supplements to regulatory and economic policy instruments.
- *Making Climate Forecasts Matter* (NRC, 1999b) developed a conceptual base and identified key scientific questions for analyzing the human consequences of seasonal-to-interannual climate variations (e.g., El Niño) and learning how make improved climate forecasting skill more useful.
- *Our Common Journey: A Transition Toward Sustainability* (NRC, 1999d) drew on nearly 375 earlier NRC reports and many other sources to develop a conceptual framework and a set of research priorities for sustainability science.
- *People and Pixels: Linking Remote Sensing and Social Science* (NRC, 1998) identified and discussed opportunities for using remotely sensed data in research on human–environment interactions and in social science, presented examples, and developed a Web-based guide to information resources.
- *Environmentally Significant Consumption: Research Directions* (NRC, 1997) conceptualized the link between consumption and environment and identified and illustrated promising research possibilities on the causes of environmentally significant consumption.

Publications primarily identifying research directions include:

- *Decision Making for the Environment* (NRC, 2005a) identified five areas of high-priority research that can contribute to improved decisions affecting environmental quality.
- *Population, Land Use, and Environment: Research Directions* (NRC, 2005b) reviewed knowledge on interactions between demo-

graphic and environmental changes mediated by land use and recommended research directions in this area.

 • *Implementing Climate and Global Change Research* (NRC, 2004) reviewed the strategic plan of the CCSP and identified areas needing additional research investment, including human dimensions, economics, adaptation, and mitigation.

 • *Human Interactions with the Carbon Cycle: Summary of a Workshop* (NRC, 2002a) reported on discussions of promising research issues linking social science and natural science analyses of the carbon cycle.

 • *Grand Challenges in Environmental Sciences* (NRC, 2001) identified eight major scientific challenges, three of which prominently featured human systems.

 • *Human Dimensions of Global Environmental Change: Research Pathways for the Next Decade* (NRC, 1999a) presented a state-of-the-field review and set of research imperatives.

 • *Research Needs and Modes of Support for the Human Dimensions of Global Change* (NRC, 1994a) recommended that NSF support a collection of centers and research teams.

 • *Science Priorities for the Human Dimensions of Global Change (*NRC, 1994b) advised NSF on the creation of a policy science program to deal with global change issues.

 • *Global Environmental Change: Understanding the Human Dimensions* (NRC, 1992) helped define human dimensions research as a coherent intellectual enterprise and recommended a plan for national research in the area.

Since the publication of *Our Common Journey* (NRC, 1999d), further statements of the fundamental research needs and priorities in sustainability science have continued to appear (e.g., Kates et al., 2001; Clark and Dickson, 2003). Members of the NRC Roundtable on Science and Technology for Sustainability and the Forum on Science and Innovation for Sustainable Development[2] have been among the sources of research priorities (see also Swart et al., 2002, on critical challenges for sustainability science, and a special issue of *PNAS*, July 8, 2003, on science in support of sustainability).

---

[2] *http://sustainabilityscience.org.*

History of Connections with Core Social Science Disciplines

Challenges in connecting human–environmental research with core disciplines in the behavioral and social sciences have been an ongoing issue for CHDGC throughout its history, reflected in both committee member appointments and in meeting agendas. For instance, most recently in cooperation with the Social, Behavioral, and Economic Sciences Directorate of the NSF, the CHDGC held a half-day symposium on April 25, 2007, on linking environmental research and the behavioral and social sciences. The initial question posed to symposium participants was: *What are the core theoretical issues that would motivate social, behavioral, and economic research on environmental topics, resulting in improved understanding of environmental phenomena as well as contributions to the core social science fields?*

The symposium included committee members and staff, participants from federal agencies, and speakers with ties to six different social and behavioral science disciplines who spoke about developments in those disciplines: political science, sociology, economics, psychology, anthropology, and geography. An underlying question from NSF was why so few social scientists submit proposals to cross-disciplinary programs related to human aspects of environmental issues. Particular problems are perceived in several social and behavioral science disciplines in which academic reward systems emphasize contributions to established core subfields or theoretical debates rather than to fundamental understanding of societal problems.

Disciplinary contributors noted significant obstacles in sociology, psychology, and political science and a split between "ecological" and "environmental" economics; environment is closer to the disciplinary core in anthropology and geography. In general, the discussion suggested that involvement by disciplinary social and behavioral scientists can be affected by the agendas and review practices of agencies such as NSF that provide major research support, especially to early-career scientists for whom an NSF grant can be an important career building block. NSF funding criteria and practices, including the composition of panels that review proposals, can help turn early-career scientists toward a focus either on established "core" disciplinary questions or on fundamental cross-disciplinary questions.

History of Interest Among CCSP Mission Agencies

Historically, except for NSF and a few other isolated programs (e.g., in the health sciences), CCSP agencies have not considered investments in fundamental human systems/human dimensions research to be a part of their mission. NSF has supported such research, though usually within broader programs (e.g., decision making under uncertainty and human and social dynamics) in which climate-related research competes with other research that is not motivated by the problems of climate change. In the other CCSP agencies, research that draws on the social sciences is mainly addressed to fairly narrow applications of science to problem solving in such mission-defined fields as environmental regulation, coastal and water resource management, agricultural and forest resource management, and energy supply and use. These programs make use of human dimensions research knowledge and tools, such as environmental economics, but they seldom invest in improving the fundamental knowledge on which such applications stand. In fact, in most cases, the more fundamental a human dimensions research question sounds, the less likely it has been to attract interest from a CCSP agency other than NSF.

In the federal environmental and energy mission agencies, none has more than limited expertise in a few fields of social science. Consequently, even if such agencies were to decide to support fundamental research on the human dimensions of issues within their purview, it would take them time to develop the staff expertise to set priorities, solicit research, set up review panels, and make full use of research results. This situation raises questions about the likelihood that the fundamental research priorities identified below will be considered relevant to CCSP agencies and program managers or, if they are considered relevant, whether they would be developed effectively. This issue is discussed in greater detail in the last section of this paper.

FUNDAMENTAL HUMAN DIMENSIONS
RESEARCH PRIORITIES

This section identifies five top substantive research priorities for fundamental research on human dimensions of climate change and three critical crosscutting research priorities. It then discusses how they were arrived at and likely benefits of investments in research on them.

Substantive Research Priorities

1. *Improving the understanding of environmentally significant consumption.* For a decade or more, the human dimensions/sustainability science communities have been saying that the single biggest weakness in the knowledge base underlying responses to climate change is a lack of understanding about human consumption linked to resource use (e.g., NRC, 1997, 1999a, 2005a; Kates, 2000). Research on environmental consumption aims to illuminate a fundamental human driver of climate change and to build understanding needed for effective mitigation responses. Part of the research agenda concerns understanding individual- and household-level behavior (e.g., what motivates consumption; links among economic consumption, resource consumption, and human well-being, including the potential to satisfy basic needs and other demands with significantly less resource consumption; and the responsiveness of consumption behavior to efforts to change it through information, persuasion, incentives, and regulations). Another part of the research agenda concerns decisions in business organizations that affect environmental resource consumption, whether by the organizations themselves, by marketing to ultimate consumers, or through the structure of product and service chains.

2. *Improving fundamental understanding of risk-related judgment and decision making under uncertainty.* Human response to climate change depends fundamentally on judgment and decision making under uncertainty, and improved fundamental understanding of these processes continues to be central to the human dimensions research agenda (e.g., NRC, 1992, 1999a, 2005a). Anticipating or guiding human systems responses to both per-

ceived risks and opportunities related to climate change and its experienced and expected impacts requires a sophisticated understanding of how people and organizations comprehend incomplete and uncertain scientific information and incorporate, ignore, or reinterpret it in decision making. The argument recently offered that advances in climate science are inherently incapable of doing much to improve the predictability of the probability of large temperature changes (Roe and Baker, 2007) helps to underline the need for increased scientific attention to understanding and improving human capacity to make wise decisions under significant and continuing uncertainty. The research agenda includes both attention to individual cognition and to risk judgments and decision making in groups, organizations, and social institutions.

3. *Improving the understanding of how social institutions affect resource use.* This topic was identified as one of eight grand challenges in environmental science (NRC, 2001) and has been repeatedly identified as a top-priority area of human dimensions research (e.g., NRC, 1999a; 2005a). The challenge is to understand how human use of natural resources is shaped by "markets, governments, international treaties, and formal and informal sets of institutions that are established to govern resource extraction, waste disposal, and other environmentally important activities" (NRC, 2001:4). Institutions create contexts and rules that shape the human activities that drive climate change and that shape the realistic possibilities for mitigation and adaptation. The research agenda includes documenting the institutions shaping these activities from local to global levels, understanding the conditions under which the institutions can effectively advance mitigation and adaptation goals, and improving the understanding of conditions for institutional innovation and change. This area has a long history in human dimensions research (see NRC, 2002c), and a relatively good scientific infrastructure, but the research questions still require considerably expanded efforts. For example, as noted in a recent special section of PNAS (Ostrom et al., 2007), many policy analysts still believe, despite considerable evidence to the contrary, that global environmental problems can be solved by a single governance system such as privatization, government control, or community control. Fundamental research on resource institutions

holds the promise of identifying more realistic behavioral models for designing responses to climate change.

4. ***Improving the understanding of socioeconomic change as context for climate change impacts and responses***. Assessing possible human systems impacts of and responses to climate change calls for an understanding of changes in other driving forces affecting those systems over the time horizon of interest in future climates. Examples include demographic change, economic change, and institutional change. Two cases are especially high priorities: technological change and land-use change.

a. One of the most significant and most difficult of socioeconomic changes to project beyond a period of one or two decades is *technological change*, which may or may not reduce the rate of climate change, reduce some of its impacts, and offer alternatives for adaptation to those impacts. The topic consistently appears on the short list of human dimensions research priorities (e.g., NRC, 1992, 1999a). Key practical applications of such research include projecting the rate of implementation of technologies for carbon capture and sequestration, affordable seawater desalination, much more efficient cooling technologies for buildings, and so forth, and finding ways to speed implementation of desired technologies. Fundamental research seeks improved understanding of what determines rates of technological innovation and adoption. The research agenda includes studies of the roles of incentives (induced technological change), of aspects of organizations that might develop and implement new technology, institutional forces promoting and resisting change, and the potential of both transformational and incremental change (e.g., historical experience with "waves of innovation").

b. A second kind of change, often a key in connecting human dimensions with Earth-system modeling, is *land-use change*, which reflects interactions between human and natural systems dimensions. This is such a central issue for climate change—related to greenhouse gas emissions, emission sinks, impacts, and responses—that it seems remarkable that a capacity does not exist to project such changes beyond a decade or two. Largely because of limitations in the ability to project demographic and economic changes over a period of more than several decades, especially at a relatively small scale, along with changes in institutional and policy contexts, however, projections of land-use change into the midterm and further are essentially unavailable at present. Needed research includes decomposing component factors influencing land-use change; improving fundamental understanding of the relationships among population, land-use change, and environment; and linkages across scales (NRC, 1998; 2005b).

5. *Valuation of climate consequences and policy responses.* No challenge is more profound in climate change mitigation, impact assessment, and response evaluation than valuing costs and benefits. To be balanced and comprehensive, judgments must confront multiple dimensions (e.g., dollars, species, and lives), multiple scales (global, regional, and local), multiple time periods, and multiple affected parties. Currently available theoretical constructs, tools, and databases are painfully inadequate for meeting this challenge. The research agenda (NRC, 1992, 2005a) includes efforts to improve the validity of formal techniques (e.g., benefit-cost analysis, contingent valuation methods) for choices in which relevant information is uncertain, in dispute, or unknown and in which the benefits and the costs go to different parties. A major emerging issue for formal analysis concerns the dynamic links and feedbacks between climate change mitigation and adaptation. For example, the costs and benefits of adaptation depend on the outcomes of efforts at mitigation, and the dependencies increase with the timescale of the analysis. The research agenda also includes efforts to design and test social processes for evaluating options (e.g., citizen juries, negotiations, public participation mechanisms) and to find ways to integrate formal scientific techniques with such processes in what have been called analytic-deliberative processes (NRC, 1996). A forthcoming NRC report on public participation in environmental assessment and decision making will elaborate on the key questions for research.

Crosscutting Priorities

1. *Observations, indicators, and metrics.* Discussions of the observational system for climate change science rarely consider the state of observations of the human systems that drive and are affected by climate change. The CCSP strategic plan and the program itself give extensive consideration to observing and monitoring states of the climate and related environmental systems, but no explicit attention to observing or monitoring human pressures on those systems or human responses to climate change. This helps to explain why data on the human component of the human–climate system are commonly recognized to be inadequate and poorly linked to data on the physical and biological components of the system (e.g., NRC, 1992, 1999a, 2005a, 2007a). As noted by participants at a CHDGC seminar in 2006 on "Human Dimensions in Major Environmental Observational Systems," although some of these observational systems include human systems variables, they rarely use systematic approaches to deciding what data to collect or how to coordinate among observational systems to enable integrated global analysis. Federal agencies that collect human systems data rarely do so in ways that allow linking to natural systems data for climate analysis. For example, the Department of Energy's data on energy consumers in households and the commercial sector are not organized so as to be useful for modeling and explaining trends in greenhouse gas emissions, and are not even considered part of the CCSP. This example can be multiplied across other federal agencies that collect data on human actions that drive climate change and that affect human vulnerability to it. Moreover, social data are typically collected in ways (e.g., subdivided by political units or nongeographical social categories) that make it difficult to link them to environmental data, for example, with geographic information systems. The CCSP would be well advised to pay systematic attention to identifying key human dimensions indicators for climate change science, identifying observational needs for developing the indicators, and developing strategies for linking human systems indicators with physical and biological indicators to enable major advances in the quantitative analysis of human–climate interactions (NRC, 2005a). The present state of the observational sys-

tem imposes severe limitations on the ability to measure and monitor vulnerability to climate change related to social and economic factors or the adaptive capacity of different regions, sectors, or populations to different kinds of climate-driven events.

2. ***Nonlinearities, feedbacks, and thresholds in system responses to climate change in a multicausal setting***. Human and natural systems are coupled in complex ways that are only beginning to be understood (e.g., Liu et al., 2007a). For example, impacts attributed to climate change are not caused by climate change alone: Most of the affected physical, ecological, social, and economic systems are simultaneously affected by a variety of human activities (renewable resource use, infrastructure development in vulnerable areas, emission of air and water pollutants, etc.) that change at least as much on a generational timescale as climate does. Moreover, efforts at mitigation affect the need for adaptation, and vice versa. To understand what to expect from climate change therefore requires understanding of the ways critical systems that support human well-being are affected by multiple stresses (NRC, 2007b) and by human activities in response to expected environmental change. Also critical is improved understanding of nonlinear dynamics, threshold effects, and the possibility of "tipping points" that shift systems into previously unknown states. Understanding such possibilities is a major challenge for science, but rapidly expanding computational capacities, combined with improved collection of data on both the human and environmental aspects of linked systems may enable new and productive kinds of modeling and analysis.

3. ***Scale dependencies and cross-scale interactions***. Issues of geographic and temporal scale pervade climate change science and policy. For example, the effects of national policies for mitigation depend on how they affect smaller units that must implement them and how they relate to policies in other countries. On the consequences side, climate change science leaders are reminded at every national workshop that most of these issues are linked inextricably with regions and locations. Climate modelers are urged to "downscale," while researchers assimilating sets of local case studies seek to "upscale." In fact, place-based approaches to integrated understanding are fundamental to sustainability science (Kates et

al., 2001; Turner et al., 2003). Yet the science base is relatively weak for understanding how human systems impacts of climate change vary across scales and how they reflect interactions among scales (e.g., Lebel and Wilbanks, 2003; NRC, 2006; Reid et al., 2006). Research needs that have been identified but not yet met include developing a bottom-up paradigm to meet the prevailing top-down paradigm for understanding climate impacts, developing a protocol for local case studies to increase the comparability of such studies, and improving the monitoring of local and small-region human systems data related to climate change impacts and responses (Wilbanks and Kates, 1999; Wilbanks, 2003).

How the Fundamental Research Priorities Were Determined

A wide range of possible research needs was identified by the earlier agenda-setting reports for global environmental change research and sustainability science already discussed. At the committee's November 2007 meeting, CHDGC members discussed the assigned question in the light of these past efforts and invited comments from distinguished committee alumni as well as other meeting participants. The resulting list of substantive research needs was reduced by the authors to the list above by informal consideration of the following issues:

• Importance of the research area in terms of climate change drivers, impacts, and responses
• Relevance of the fundamental research across multiple applied research areas
• Potential for connecting with and drawing on core disciplinary strengths
• Potential for payoff in decision support
• Readiness of the scientific community to make progress in the area

Time did not allow for more formal or systematic efforts to select among possible priorities. Continued discussion with a broader cross section of researchers is likely to refine this list.

Potential Benefits from the Research

Taken together, the above research priorities make up a program of activities that would advance fundamental understanding of the human systems factors that drive climate change and that shape the human capacity to respond. Research on priorities 1, 3, 4, and 6 would improve basic understanding of the human forces driving climate change, and research on these priorities and priorities 7 and 8 would improve the ability to model and forecast these human drivers. Taken together with research on the natural systems aspects of climate science, this research would improve the ability to project the human impacts of climate change. Research on priorities 2, 4, and 6 would help improve the ability of individuals and decision-making organizations to gain a more complete understanding of the implications of response options and thus to make better informed and more widely accepted choices.

There are also potential benefits in terms of problem-specific knowledge and agencies' mission responsibilities. Many of the above priorities can be pursued by individual mission agencies in their particular contexts (e.g., fundamental research on consumption can be carried out in the arenas of energy use, development of coastal lands, and air and water pollution). Such efforts can contribute to the mission goals of the sponsoring agency and also to the development of basic knowledge that can be extended to benefit other agencies. This pattern of linking specific to general knowledge is illustrated by the development of knowledge about institutions and environmental resource use (priority 3). Research on this topic has been sponsored by agencies responsible for managing forests, fisheries, and international development, as well as by private foundations. Drawing on knowledge from varied research contexts, an international network of researchers has been building a body of fundamental knowledge that is providing useful insights and principles applicable in new contexts (NRC, 2002c; Dietz et al., 2003).

PRIORITIES FOR ACTION-ORIENTED
HUMAN DIMENSIONS RESEARCH

Human dimensions research inspired by the challenges of climate change can be placed along a continuum from fundamental research to targeted, focused, or action-oriented research. At the fundamental end of the continuum lies research to understand the most basic phenomena underlying human interactions with the climate system: environmentally significant consumption, risk-related judgment and decision making, and the other topics identified in the preceding section. Such research may not examine climate issues directly, but it illuminates processes that fundamentally shape human interactions with climate. As already noted, fundamental human dimensions research is sometimes not recognized by government agencies as mission relevant.

Near the targeted or focused end of the continuum lie research activities addressed to specific climate response issues, often of obvious relevance to mission agencies. Such research might inform the design or implementation of policies or the pursuit of specific priorities of government agencies or private-sector organizations. For example, a government agency with a mission to mitigate greenhouse gas emissions might commission research to determine the most effective way to inform builders, mortgage lenders, and homeowners about the energy efficiency of buildings, with the goal of facilitating their decisions. The research might compare the effects of a certification system such as the Energy Star program for appliances with a rating system and with labels that provide numerical measures in energy or carbon units or with a monetary metric, such as energy cost of ownership. An emergency management agency might commission research to determine the readiness of first responders in a city to deal with a major coastal storm or a rush to hospital emergency rooms caused by a heat wave. Examples at this level of specificity could obviously be listed almost ad infinitum.

Priority setting for research at the most focused end of the continuum makes sense only within the context of a specific mission. This section of the paper therefore focuses on a level of specificity between the ends of the continuum, where it makes sense to identify priorities at the level of the CCSP. This was done by considering some of the main areas of CCSP responsibility, examining NRC reports and other key sources in these areas, and developing the following list of focused or action-oriented human dimensions research priorities. As there were no systematic deliberative processes to consult that developed priorities at this level, the list is presented very tentatively as a basis for further discussion.

1. *Understanding climate change vulnerabilities: Human development scenarios for potentially affected regions, populations, and sectors.* The impacts of climate change depend on the conjunction of physical and biological events, driven by climatic processes, with social and economic developments occurring on the same timescales in the affected places. Much attention has been given to improving projections of future biophysical events, but far less has been given to measuring and projecting the social, economic, and cultural conditions that determine the human consequences of those events: the ways economic development, human population dynamics, investments in physical infrastructure and emergency response capabilities, changes in the demand for water and other resources, land-use change, emissions of toxic substances, and other changes combine to alter the populations, places, and sectors that may experience climate-related shocks and thus affect their vulnerabilities. Research is needed to gather and organize data on these social forces and to build methods and models for estimating, analyzing, and projecting human vulnerabilities to climate change. The absence of past efforts to build linked time-series databases covering these variables is an impediment to progress, but there are useful data in many parts of the world that could be linked. The research could examine vulnerability on several dimensions: by type of climate-driven event (storm surge, crop failure, heat wave, changing ecology of disease, etc.), by location and scale, by socioeconomic characteristics of affected populations, and by sector (market and subsistence agriculture, water supply and quality, insured and uninsured property, public health, etc.). Estimates of the time trajectories of these vulnerabilities would yield scenarios of vulnerability that could be integrated

with climate scenarios to produce improved projections of the impacts of climate change (NRC, 1998, 1999b, 2008).

2. *Understanding mitigation potential: Driving forces, capacities for change, and possible limits of change*. Discussions of climate change mitigation are more often rooted in policy targets and integrated assessment modeling than in solid, evidence-based studies of the behavior of individuals, organizations, and economies. We know that highly aggregated models of some of the drivers of climate change, such as energy and land use, have often been far off the mark in predicting future trends. Building such models from disaggregated data on population dynamics, economic activity, energy and resource demand, and other social indicators has the potential to yield improved forecasts based on better understanding of the underlying processes (NRC, 1984, 1992, 1997, 1998, 2005b). For example, carbon emissions per capita decreased by 10 percent in the United States between 1973 and 2001, but there was considerable variation between neighboring states: Per capita carbon emissions decreased in California by 31 percent in that period while they increased by 2 percent in Arizona; they decreased in Minnesota by 14 percent while they increased by 29 percent in Iowa (Blasing et al., 2004). Understanding such differences can build deeper understanding of ongoing changes in carbon emissions and how they respond to various forces in the economic, social, and policy environments.

Efforts to mitigate climate change by altering the driving forces depend on inducing social and behavioral change in individuals, organizations, and institutions. Much of the needed change takes the form of inducing innovation and adoption of technologies for energy efficiency and low-carbon energy production and for the design of communities and other physical infrastructure; some involves changes in the use of existing technology and infrastructure. Change can potentially be accomplished by various combinations of regulatory action, standard setting, information, financial incentives, and voluntary action. However, research is needed to find the right combinations and to assess the efficacy of policy alternatives. The effects of particular interventions, such as providing financial incentives, are sometimes much less than expected, and highly variable depending on the target actors and how the policies are implemented (NRC, 1984, 1985, 1997, 2005a). Various NRC reports have elaborated on segments of the research agenda, for example, by reviewing knowledge on the potential of education, information, and voluntary measures (NRC, 2002b); the effects of tradable-permit and community-based management approaches (NRC, 2002c); and mitigation-potential issues in sectors such as households (NRC, 1985, 2005a) and businesses (NRC, 2005a).

3. *Understanding adaptation contexts, capacities for change, and possible limits of change.* Adaptation to climate change is a matter of how regions, sectors, populations, and their governing institutions cope with their vulnerabilities (Adger et al., 2007). It is a matter of anticipation, anticipatory actions to reduce vulnerability, immediate responses to climate-related events, and recovery, and includes actions within various risk management systems (e.g., physical infrastructure, emergency response systems, insurance). Adaptive capacity varies with the type of event, the place, the time frame, and attributes of the affected human systems (e.g., Turner et al., 2003; Smit and Wandel, 2006). There are several research needs. One is to develop indicators of adaptive capacity that can address the diversity of types of disruptive events; assess effects by region, sector, human activity, and timescale; incorporate assessments of coping capacity (e.g., emergency preparedness and response systems, insurance systems, disaster relief capabilities); and consider diverse types of impacts (e.g., on life and health, economic systems, business organizations, governments, and communities; see Yohe and Tol, 2002; Brooks and Adger, 2005). Another need is to assess various generic and event-specific adaptation options in terms of their ability to reduce unwanted consequences of climate change. Like indicators, these assessments depend on timescale. For example, levees can protect against coastal storms expected on a timescale of decades, but on a timescale of centuries, urban redesign or relocation may yield better results. Yet another is to assess barriers to adaptation, which can be significant even when capacity to adapt is high (Adger et al., 2007). These lines of research will lead to (a) more comprehensive models of the effects of climate change that take into account vulnerabilities, resilience, and coping responses; (b) improved means to prepare for effects and improve resilience;

and (c) better informed public debate about the trade-offs involved in coping with the threats posed by climate change.

4. ***Understanding how mitigation and adaptation combine in determining human systems risks, vulnerabilities, and response challenges associated with climate change.*** Along with the importance of improving the scientific understanding of mitigation and adaptation as separate research priorities, a rapidly emerging need is to improve the ability to consider mitigation and adaptation as joint contributors to an integrated approach to climate change responses (Wilbanks et al., 2003; Klein et al., 2007). It is clear that both are needed, as mitigation seeks to keep climate change to a level at which adaptation can cope with most of the impacts, and as adaptation makes it possible to live with more realistically achievable mitigation targets. In developing its Fourth Assessment Report, the Intergovernmental Panel on Climate Change made an effort to overcome organizational constraints to address such integration issues in its Working Group II report, based partly on an "Expert Meeting on Integration of Adaptation, Mitigation, and Sustainable Development" in La Reunion, February 2005. But, with limited resources to understand the relationships of mitigation and adaptation, it is virtually impossible at present to analyze questions of balance, possible complementarities and cobenefits, and relationships with sustainable development. The research challenges include filling gaps in information needed for analysis and addressing the differences between mitigation and adaptation options in their character (how, where, and when they work), their agency (who decides), and who benefits and who pays. Moreover, the two sets of research communities tend to be divided by a vast gulf of different methodologies and practices. The benefits include a more realistic and comprehensive understanding of climate response options, their relationships to each other, and their joint effects on the human consequences of climate change.

5. ***Understanding decision support needs for climate change responses and how to meet them.*** The success of efforts to develop a national climate service or the like will depend on its ability to provide credible, timely, and decision-relevant information to its constituencies. Research is needed to understand, for various classes of decision makers: the kinds of climate information that could help

them make better resource management and adaptation decisions, the ways such information can be made to fit into their decision routines, the factors that determine whether potentially useful information is actually used, and the forms and sources of information that would make it most useful (NRC, 1999b, 2005a, 2008). This research should seek to improve the match between what science can provide and what decision makers need by identifying scientific information that would add value for users, finding better ways to deliver that information, and finding better ways to incorporate users' needs into research agenda-setting processes.

This priority includes two related but distinct elements. One involves research to improve institutions for communication, such as networks that link the producers and users of information, usually through intermediary individuals or organizations (NRC, 2008). This research can help speed the evolution of effective networks by allowing them to build on basic social science knowledge and evaluations of past experience. The other element relates to the development of decision "tools," messages, and other products that convey important information from its producers to intermediaries and from them to ultimate users. Research to develop networks and tools is clearly distinct from their actual operation, which can be considered to be an outreach or extension activity (NRC, 2008). The conflation of research and operational decision support activities in CCSP documents makes it difficult to assess the research elements of the program (NRC, 2007a); in our view, investments in outreach should not be counted as research.

Networks and tools are needed for both mitigation and adaptation. On the mitigation side, for example, businesses, governments, and households need tools to assess their emissions and evaluate ways to reduce them. Many available instruments, such as carbon calculators, are neither reliable nor transparent (Padgett et al., 2008). Research can help develop communication networks that deliver such information in ways that users find credible, salient, and understandable (NRC, 1984, 1989, 2002b). Adaptation also requires development of decision support tools and networks, though the tools have different purposes (e.g., estimating the probability and severity of various kinds of climate-related extreme events), and the networks must link different sets of information providers and users.

6. ***Coordinating response efforts across scales***. Both mitigation and adaptation responses will depend on coordination among actors that function at different social and geographical scales. For mitigation, a key issue is the need for international coordination: National policies have little effect if they run counter to what is happening in other countries. There have been past successes, as with ozone-depleting substances, but climate mitigation is an especially hard case because the emitters are so diffuse. Coordination must reach actors below the national level, either via national policy or in other ways (e.g., international voluntary agreements and standards; Prakash and Potoski, 2006). For adaptation, a key issue is that many of the affected localities, sectors, and populations lack the scientific resources needed to anticipate impacts and the financial resources needed to invest in anticipatory responses. These resources are likely to be available only from higher level actors, who must in turn develop ways to provide useful information to lower level actors, identify and prioritize needs, and coordinate responses (Wilbanks, 2007). Although there is a knowledge base from past experiences with cross-scale coordination, climate change provides new challenges, including the importance of global science for local adaptation. Pieces of the research agenda have been developed in several NRC studies (e.g., NRC, 2002b, c, 2008).

Many of the fundamental research priorities outlined in the preceding section are directly related to these more focused research needs by way of providing the scientific underpinning for them. Table D.1 shows how the two kinds of research are in fact closely related. For each of the six action-oriented research priorities, it identifies the fundamental research priorities that are relevant to the more focused need. Circles indicate the especially strong links.

CRITICAL CONSTRAINTS ON PROGRESS

The continuing underinvestment in human dimensions research (and especially in fundamental human dimensions research), since it was first noted in NRC reviews in 1990, suggests that a lack of well-defined research priorities is not the main barrier to progress. This section discusses four additional fundamental barriers that have been identified in past NRC reports (NRC, 1990, 1992, 1999a, 2004, 2005a, 2007a), beginning with the ones that are most straightforward to address and moving to the most challenging ones; and, as a basis for further discussion, it offers some ideas based on these reports.

1. *Limitations in total level of support.* The initial NRC review of the U.S. Global Change Research Program (NRC, 1990) concluded that the human interactions science priority, then at 3 percent of the budget, was "the most critically underfunded in the fiscal 1991 budget for the USGCRP." A 1992 NRC report recommended that the level of support for human dimensions research be increased from 3 percent of the program (in FY 1991) to 5 percent, with a 3-year ramp-up period. This level was considered justified not only by need, but also by the capability of the research community to use the increased funds effectively. The report offered a set of concrete recommendations on how to allocate additional funds (approximately $20 million per year): an additional $7 million per year for new investigator-initiated basic and targeted research, $10 million for 100 new 2-year postdoctoral fellowships per year, and an increment of $2 million to $4 million for data acquisition and dissemination. The report noted that the data issue in particular could benefit from even larger increases in funding. In 2004, the NRC review of the CCSP strategic plan noted the need to accelerate efforts in "underemphasized program elements," including "human dimensions, economics, impacts, mitigation, and adaptation" (NRC, 2004:2). By FY 2006, however, human dimensions funding had declined from the FY 1991 level to the point that the relevant program element, which included both human dimensions research and an undetermined amount of nonresearch expenditures on decision support systems, constituted 2 percent of the program (NRC, 2007a). The most recent NRC review (NRC, 2007a:4) concluded that "Progress in human dimensions research has lagged progress in natural climate science . . . . This disparity in progress likely reflects the inability of the CCSP to support a consistent and cogent research agenda as recommended in previous studies."

TABLE D.1 Relationship Between Fundamental Research Priorities and Focused Research Needs

Fundamental Research Priorities	Action-Oriented Human Dimensions Research Priorities					
	Vulnerability Scenarios	Mitigation Potentials	Adaptation Potentials	Integrating Mitigation and Adaptation	Effective Decision Support	Responses Across Scales
Consumption	X	(X)	X	X	X	X
Risk-related decision making	X	X	X	X	(X)	X
How social institutions affect resource use	X	X	X	X	(X)	X
Socioeconomic changes as contexts for impacts and responses	(X)	X	X	X	X	X
Valuation of consequences and responses	X	X	X	(X)	X	X
Observations, indicators, and metrics	(X)	(X)	(X)			(X)
Nonlinearities, feedbacks, and thresholds in a multicausal setting	X	X	(X)	X		X
Scale dependence and interactions	X	X	X	X	X	(X)

Given increasing demand in recent years for human dimensions knowledge to inform responses to climate change, we believe that a level of support beyond 5 percent of the total program could be justified today. Determining how much of an increase can be effectively absorbed by the research community, and how fast, will require more careful analysis of capabilities than is possible with available information.

Because of the longstanding underfunding of human dimensions research, we do not think that expanded development of the above priority areas can be achieved by reallocating funds from other areas of human dimensions research that are already well developed. If funding for climate change research continues to be flat, we think these priorities can only be developed by reallocating funds from well-developed areas of natural systems research. The needed reallocation may be small, however, because human systems research is typically far less expensive than natural systems research.

2. *Data needs and limitations*. As mentioned previously, the shortage of human systems data in forms useful for analysis of human dimensions issues has been discussed in a number of NRC reports (e.g., NRC, 1992, 1999a, 2005a). There is a particular need for time-series data regarding human pressures on the global environment, such as data on land cover and land use, extraction of natural resources from ecosystems, energy consumption and pollutant emissions from various sources and sectors, human attitudes, valuations, and responses. There is a similar need for data on human consequences of and responses to global environmental change, such as morbidity and mortality data related to air and water quality and vulnerabilities to climate-related extreme events. A major constraint on progress in modeling and understanding human–climate interactions is the lack of reliable, linked databases tracking human activities and the natural systems they change and that in turn affect them. The NRC recommended in 1992 that the first steps include efforts to identify major data needs, inventory available datasets from public and private sources, and assess what would be needed to (a) link existing human and environmental datasets and (b) fill critical gaps in existing data bases. At that time, it recommended that an extra $2 million to $4 million per

year be devoted to data acquisition and dissemination. Subsequent NRC reports (NRC, 1999c, 2005a) have expanded on these recommendations to identify types of indicators that should be developed to improve understanding of human interactions with the global environment. The most recent NRC review of the CCSP (NRC, 2007a:79) noted an institutional difficulty related to the data needs: "[T]he need to collect social, economic, and health data to address the human dimensions aspects of the program adds an additional level of complexity because these data are outside the purview of agencies traditionally associated with climate measurements." We do not think it is an exaggeration to say that in most of the CCSP agencies, the concept of observations does not include observations of human activities or human conditions. A comprehensive approach to addressing the data constraint would be a major effort to *develop linked data* on social and environmental phenomena in time series, across space, and at multiple analytical levels. A human dimensions observational system would complement the natural science observational systems that are so central to the CCSP. Such an effort, in addition to its value for research, could provide new opportunities for social science research that would attract early-career and established researchers from the disciplines into global change research and thus expand the pool of strong researchers in the area. This approach also could help encourage agencies to support fundamental human dimensions research by demonstrating the mission relevance of research using social science concepts and variables.

3. ***Connections with the basic social and behavioral sciences***. As already noted, fundamental human dimensions research pursues questions driven by concerns with problems of climate change rather than issues arising from the social science disciplines. There are connections between the two, but they are not always obvious to researchers in the disciplines. Moreover, the relevance of disciplinary concepts to climate problems does not always seem the same from a disciplinary standpoint as from a climate problem perspective. For example, household energy consumption is an important contributor to global greenhouse gas emissions, but understanding it requires concepts from multiple disciplines. Efforts to explain it only in terms of environmental attitudes (psychology), social position

(sociology), or household income (economics) are likely to seem naïve or seriously incomplete to scientists who take a broader view of the climate problem. The problem of linking the disciplines to climate questions is in part one of developing theory and method. Issues such as environmental consumption, land-use change, and valuation of environmental resources, among others, do not yield easily to discipline-specific concepts, theories, or methods. Arguably, multidisciplinary approaches are more likely to yield useful tools for answering questions about human–climate interactions. The roles of disciplinary tools must be worked out over time in research teams and the wider community. Without sufficient support for such teams to work together over time, progress will be restrained.

The problem is also one of human resource development. Scientists with a strong social science disciplinary background have a learning curve to traverse before they can make serious contributions to understanding the climate problem. In our view, so far the CCSP as a program has not made efforts to speed this learning, for example, by supporting interdisciplinary graduate training programs in climate science that encourage social scientists to apply. More could be done to draw social scientists to climate change research, particularly at the predoctoral and early career stages, as noted in past NRC reviews. The 1992 NRC review devoted a chapter to human resource and organization issues and offered several recommendations for addressing the problem, including the creation of a transportable 5-year package of dissertation, postdoctoral, and research support. This idea was echoed in a proposal voiced at the CHDGC's 2007 workshop on linking environmental research and the social and behavioral sciences. The idea would be to facilitate career advancement for social scientists working in a field outside the core of their disciplines, which could help build the community of researchers and might strengthen interdisciplinary institutions working on climate change. As the most recent NRC (2007a:92) review noted, "The natural sciences may offer a successful model for building human dimensions capacity, especially programs to move young investigators into the arena and to support postdocs."

4. *Organizational barriers in the federal government*. The 1992 NRC review concluded that there was "an almost complete

mismatch between the roster of agencies that support research on global change and the roster of agencies with strong capabilities in social science" (NRC, 1992:10). As already noted, NSF is essentially the only CCSP agency in which fundamental research in the human systems sciences is considered part of the agency's mission. Climate change, however, is not central to the NSF mission. CCSP agencies with climate missions seem ready to recognize fundamental natural systems research related to climate change as falling within their missions, but much less ready to accord the same recognition to human systems research. Some of these agencies support human dimensions research in particular applied areas, some of it quite valuable to the CCSP, but in our view these efforts have done little to build the kinds of fundamental knowledge prioritized earlier in this paper. The CCSP and its agencies could show leadership in addressing this challenge by supporting fundamental research on the human-system components of climate change.

The 1992 study proposed that federal agencies that support basic natural science research on global change, but only applied social science, expand their support to include fundamental social science research related to specific global change topics of interest to them. This is one approach to overcoming organizational barriers in the agencies. Subsequent NRC reports have noted the lack of a programwide office with significant budgetary authority. Organizational barriers might be addressed in part by endowing the program office with sufficient authority and staffing to develop the human interactions program element centrally.

These constraints have all been present for a long time, and the research community has been aware of them for a long time. All of them were identified in the first NRC (1992) review of human dimensions in the U.S. Global Research Program. Not since that review has any NRC group attempted to define in any detail implementation strategies for overcoming them. The 1992 review, particularly Chapter 6 on data needs, Chapter 7 on human resources and organizational structures, and Chapter 8 on the structure of a national research program, still offers the most comprehensive analysis available for developing recommendations for implementing an effective human dimensions element in the CCSP. However, it requires updating to take into account progress since then in supplying human dimensions knowledge and a major increase in demand for it, occasioned by greatly increased acceptance of the need for science to understand the human consequences of climate change and to inform decisions on how to respond to the associated risks and opportunities.

Responding to these constraints also requires attention to the history of the program's responses since 1990 to NRC recommendations to expand human dimensions research. This history suggests to us that some of the organizational barriers in the CCSP and its participating agencies are strongly entrenched. We therefore suggest that the Strategic Advice Committee give serious attention to ways to overcome organizational barriers, so that its priorities for human systems research will stand a better chance of implementation than the recommendations of past NRC study committees. We suggest that the committee consider such recommendations as (a) reconsidering the purposes of the program at a time when the national concern has broadened from documenting and attributing climate change to informing responses to it, (b) undertaking new commitments at the program level, (c) making organizational changes in the climate programs of CCSP mission agencies, and (d) hiring staff in the program office and some of the agencies with the expertise and authority required to organize the needed research.

SUMMARY

Reviews of the CCSP and the previous Global Change Research Program have repeatedly found significant underinvestment in research on the human systems and their interactions with climate. Drawing on these reviews and recent discussions at meetings of the NRC Committee on the Human Dimensions of Global Change, this paper identifies five substantive research priorities for developing the human systems side of climate science: research on environmentally significant consumption, judgment and decision making under uncertainty, institutions and climate change, technological change, and valuation of climate change and human responses. Three crosscutting science priorities are also critical:

developing human systems observations, indicators, and metrics and linking them to natural systems data; understanding nonlinearities and complexities in system response; and understanding scale dependencies and interactions. The paper also identifies six priorities for research that is more action-oriented in the near term than the fundamental research that is suggested. For any of these priorities to be implemented, four critical constraints on scientific progress require attention: limited levels of support, data gaps, limited capability for multidisciplinary environmental research among researchers with social and behavioral science expertise, and perhaps most critical, organizational barriers to human dimensions research in federal agencies responsible for climate change science. The paper briefly discusses implementation issues related to overcoming these constraints.

REFERENCES

Adger, W.N., S. Agrawala, M.M.Q. Mirza, C. Conde, K. O'Brien, J. Pulhin, R. Pulwarty, B. Smit, and K. Takahashi, 2007, Assessment of adaptation practices, options, constraints and capacity, in *Climate Change 2007: Impacts, Adaptation and Vulnerability. Contribution of Working Group II to the Fourth Assessment Report of the Intergovernmental Panel on Climate Change*, M.L. Parry, O.F. Canziani, J.P. Palutikof, P.J. van der Linden, and C.E. Hanson, eds., Cambridge University Press, Cambridge, pp. 717–743.

Blasing, T.J., C.T. Broniak, and G. Marland, 2004, Estimates of annual fossil-fuel CO_2 emitted for each state in the U.S.A. and the District of Columbia for each year from 1960 through 2001, in *Trends: A Compendium of Data on Global Change*, Carbon Dioxide Information Analysis Center, Oak Ridge National Laboratory, available at *http://cdiac.esd.ornl.gov/trends/emis_mon/stateemis/emis_state.htm.*

Brooks, N., and W.N. Adger, 2005, Assessing and enhancing adaptive capacity, in *Adaptation Policy Frameworks for Climate Change*, B. Lim, E. Spanger-Siegfried, I. Burton, E.L. Malone, and S. Huq, eds., Cambridge University Press, New York, pp. 165–182.

Clark, W.C., and N.M. Dickson, 2003, Sustainability science: The emerging research program, *Proceedings of the National Academy of Sciences*, **100**, 8059–8061.

Dietz, T., E. Ostrom, and P.C. Stern, 2003, The struggle to govern the commons, *Science*, **302**, 1907–1912.

Kates, R., 2000, Population and consumption: What we know, what we need to know, *Environment*, **10**, 12–19.

Kates, R.W., W.C. Clark, R. Corell, J.M. Hall, C.C. Jaeger, I. Lowe, J.J. McCarthy, H.J. Schellnhuber, B.Bolin, N.M. Dickson, S.Faucheux, G.C. Gallopin, A. Grübler, B.Huntley, J. Jäger, N.S. Jodha, R.E. Kasperson, A. Mabogunje, P. Matson, H. Mooney, B. Moore III, T. O'Riordan, and U. Svedin, 2001, Sustainability science, *Science*, **292**, 641–642.

Klein, R., J.A. Sathaye, and T.J. Wilbanks, 2007, Challenges in integrating mitigation and adaptation as responses to climate change, special issue, *Mitigation and Adaptation Strategies for Global Change*, **12**, 639–962.

Lebel, L., and T. Wilbanks, eds., 2003, Dealing with scale, *Ecosystems and Human Well-Being: A Framework for Assessment*, Millennium Ecosystem Assessment, Island Press, Kuala Lumpur, pp. 107–126.

Liu, J., T. Dietz, S.R. Carpenter, M. Alberti, C. Folke, E. Moran, A.N. Pell, P. Deadman, T. Kratz, J. Lubchenco, E. Ostrom, Z. Ouyang, W. Provencher, C.L. Redman, S.H. Schneider, and W.W. Taylor, 2007a, Complexity of coupled human and natural systems, *Science*, **317**, 1513–1516.

Liu, J., T. Dietz, S.R. Carpenter, C. Folke, M. Alberti, C.L. Redman, S.H. Schneider, E. Ostrom, A.N. Pell, J. Lubchenco, W.W. Taylor, Z. Ouyang, P. Deadman, T. Kratz, and W. Provencher, 2007b, Coupled human and natural systems, *Ambio*, **36**, 639–648.

NRC (National Research Council), 1984, *Energy Use: The Human Dimension*, Freeman, New York, 237 pp.

NRC, 1985, *Energy Efficiency in Buildings: Behavioral Issues*, P.C. Stern, ed., National Academy Press, Washington, D.C., 110 pp.

NRC, 1989, *Improving Risk Communication*, National Academy Press, Washington, D.C., 352 pp.

NRC, 1990, *Research Strategies for the U.S. Global Change Research Program*, National Academy Press, Washington, D.C., 294 pp.

NRC, 1992, *Global Environmental Change: Understanding the Human Dimensions*, P.C. Stern, O.R. Young, and D. Druckman, eds., National Academy Press, Washington, D.C., 320 pp.

NRC, 1994a, *Research Needs and Modes of Support for the Human Dimensions of Global Change*, National Academy Press, Washington, D.C.

NRC, 1994b, *Science Priorities for the Human Dimensions of Global Change*, National Academy Press, Washington, D.C., 44 pp.

NRC, 1996, *Understanding Risk: Informing Decisions in a Democratic Society*, P.C. Stern and H.V. Fineberg, eds., National Academy Press, Washington, D.C., 264 pp.

NRC, 1997, *Environmentally Significant Consumption: Research Directions*, P.C. Stern, T. Dietz, V.W. Ruttan, R.H. Socolow, and J.L. Sweeney, eds., National Academy Press, Washington, D.C., 152 pp.

NRC, 1998, *People and Pixels: Linking Remote Sensing and Social Science*, National Academy Press, Washington, D.C., 256 pp.

NRC, 1999a, *Human Dimensions of Global Environmental Change: Research Pathways for the Next Decade*, National Academy Press, Washington, D.C., 100 pp.

NRC, 1999b, *Making Climate Forecasts Matter*, P.C. Stern and W.E. Easterling, eds., National Academy Press, Washington, D.C., 192 pp.

NRC, 1999c, *Nature's Numbers: Expanding the National Economic Accounts to Include the Environment*, W.D. Nordhaus and E.C. Kokkelenberg, eds., National Academy Press, Washington, D.C., 262 pp.

NRC, 1999d, *Our Common Journey: A Transition Toward Sustainability*, National Academy Press, Washington, D.C., 384 pp.

NRC, 2001, *Grand Challenges in Environmental Sciences*, National Academy Press, Washington, D.C., 106 pp.

NRC, 2002a, *Human Interactions with the Carbon Cycle: Summary of a Workshop*, P.C. Stern, ed., National Academy Press, Washington, D.C., 50 pp.

NRC, 2002b, *New Tools for Environmental Protection: Education, Information, and Voluntary Measures*, T. Dietz and P.C. Stern, eds., National Academy Press, Washington, D.C., 368 pp.

NRC, 2002c, *The Drama of the Commons*, E. Ostrom, T. Dietz, N. Dolsak, P.C. Stern, S. Stonich, and E.U. Weber, eds., National Academy Press, Washington, D.C., 534 pp.

NRC, 2004, *Implementing Climate and Global Change Research: A Review of the Final U.S. Climate Change Science Program Strategic Plan*, National Academies Press, Washington, D.C., 108 pp.

NRC, 2005a, *Decision Making for the Environment: Social and Behavioral Science Research Priorities*, G.D. Brewer and P.C. Stern, eds., National Academies Press, Washington, D.C., 296 pp.

NRC, 2005b, *Population, Land Use, and Environment: Research Directions*, B. Entwisle and P.C. Stern, eds., National Academies Press, Washington, D.C., 344 pp.

NRC, 2006, *Facing Hazards and Disasters: Understanding Human Dimensions*, National Academies Press, Washington, D.C., 408 pp.

NRC, 2007a, *Evaluating Progress of the U.S. Climate Change Science Program: Methods and Preliminary Results*, The National Academy Press, Washington, D.C., 178 pp.

NRC, 2007b, *Understanding Multiple Environmental Stresses: Report of a Workshop*, National Academies Press, Washington, D.C., 154 pp.

NRC, 2008, *Research and Networks for Decision Support in the NOAA Sectoral Applications Research Program*, H.M. Ingram and P.C. Stern, eds., National Academies Press, Washington, D.C., 98 pp.

Ostrom, E., M.A. Janssen, and J.M. Anderies, eds., 2007, Going beyond panaceas, *Proceedings of the National Academy of Sciences*, **104**, 15,176–15,223.

Padgett, J.P., A.C. Steinemann, J.H. Clarke, and M.P. Vandenbergh, 2008, A comparison of carbon calculators, *Environmental Impact Assessment Review*, **28**, 106–115.

Prakash, A., and M. Potoski, 2006, *The Voluntary Environmentalists: Green Clubs, ISO 14001, and Voluntary Environmental*

Regulations, Cambridge University Press, Cambridge, 198 pp.

Reid, W., F. Berkes, T. Wilbanks, and D. Capistrano, eds., 2006, *Bridging Scales and Knowledge Systems: Concepts and Applications in Ecosystem Assessment*, Island Press, Washington, D.C., 351 pp.

Roe, G.H., and M.B. Baker, 2007, Why is climate sensitivity so unpredictable? *Science*, **318**, 629–632.

Shapley, D., and R. Roy, 1985, *Lost at the Frontier: U.S. Science and Technology Policy Adrift*, ISI Press, Philadelphia, 223 pp.

Smit, B., and J. Wandel, 2006, Adaptation, adaptive capacity and vulnerability, *Global Environmental Change*, **16**, 282–292.

Stern, P.C., 1993, A second environmental science: Human-environment interactions, *Science*, **260**, 1897–1899.

Swart, R., P. Raskin, J. Robinson, R. Kates, and W.C. Clark, 2002, Critical challenges for sustainability science, *Science*, **297**, 1994–1995.

Turner, B.L., R.E. Kasperson, P.A.Matson, J.J.McCarthy, R.W. Corell, L. Christensen, N. Eckley, J.X. Kasperson, A. Luers, M.L.Martello, C. Polsky, A. Pulsipher, and A. Schiller, 2003, A framework for vulnerability analysis in sustainability science, *Proceedings of the National Academy of Sciences*, **100**, 8074–8079.

Wilbanks, T.J., 2003, Geographic scaling issues in integrated assessments of climate change, in *Scaling Issues in Integrated Assessment*, J. Rotmans and D. Rothman, eds., Swets and Zeitlinger, Lisse, The Netherlands, pp. 5–34.

Wilbanks, T.J., 2007, Scale and sustainability, *Climate Policy*, **7**, 278–287.

Wilbanks, T.J., and R. Kates, 1999, Global change in local places, *Climatic Change*, **43**, 601–628.

Wilbanks, T.J., S.M. Kane, P.N. Leiby, R.D. Perlack, C. Settle, J.F. Shogren, and J.B. Smith, 2003, Possible responses to global climate change: Integrating mitigation and adaptation, *Environment*, **45**, 28–38.

Yohe, G., and R.S.J. Tol, 2002, Indicators for social and economic coping capacity—Moving toward a working definition of adaptive capacity, *Global Environmental Change*, **12**, 25–40.

Appendix E

Research Priorities for Improving Our Understanding of the Natural Climate System and Climate Change

Antonio J. Busalacchi, University of Maryland
Ian Kraucunas, National Research Council

Note: The committee commissioned the following discussion paper from the staff and chair of the National Research Council Climate Research Committee. Their views, as expressed below, may not always reflect the views of their committee, the Committee on Strategic Advice on the U.S. Climate Change Science Program, or vice versa.

BACKGROUND

The National Academies' Committee on Strategic Advice on the U.S. Climate Change Science Program (CCSP Advisory Committee) is charged to "examine the program elements described in the Climate Change Science Program strategic plan and identify priorities to guide the future evolution of the program in the context of established scientific and societal objectives." These priorities may include "adjustments to the balance of science and applications, shifts in emphasis given to the various scientific themes, and identification of program elements not supported in the past." To help develop its response to this charge, the CCSP Advisory Committee has requested input on:

1. Top priorities that have been identified by the Climate Research Committee (CRC) and elsewhere that focus on understanding the climate system and that take into consideration science requirements from stakeholders
2. Sources/references for these priorities and criteria for selecting them
3. Items in the existing CCSP strategic plan that could be deemphasized

This discussion paper identifies 15 priorities, with an emphasis on improving our scientific understanding of the *natural* climate system. To develop this list of priorities, we reviewed the documents listed in the "Context" section below, solicited input from members of the CRC and its parent Board on Atmospheric Sciences and Climate (BASC), and used an informal set of criteria, listed in the "Selection Criteria" section, to select 15 priorities from among the many ideas collected. We identified two overarching priorities, which emerged as the most critical issues facing the CCSP from the perspective of natural sciences research, then categorized the remaining 13 as either existing priorities (i.e., those already reflected in the 2003 CCSP strategic plan), emerging priorities (those that have surfaced or increased in importance during the past 5 years), or crosscutting (infrastructural and organizational) priorities.

Although an attempt was made to cover the full range of activities needed to facilitate progress in understanding the physical basis of climate change and to support climate-related decision making, this appendix does not attempt to provide a comprehensive review of the priorities listed in other documents (e.g., the recently proposed draft revisions to the CCSP strategic plan), and it is likely that some important priority areas have been overlooked. Also, this appendix focuses on priorities related to the natural (physical-biogeochemical) climate system; a companion paper (Appendix D) discusses priorities for the human dimensions of global change research. The CCSP Advisory Committee will also be considering priorities for climate science applications, which were the focus of a workshop held in October 2007. These collections of ideas are all intended to serve as a starting point for discussions at the CCSP Advisory Committee's March 2008 workshop, where that committee will begin developing its final report on future priorities for the CCSP.

CONTEXT

Our understanding of the climate system and climate change has evolved rapidly over the past several decades. Significant progress has been made in many areas, such as measuring the precise concentrations of different greenhouse gases and determining their impact on Earth's radiative balance, while other questions have proven more challenging to answer. A number of documents produced by the National Research Council (NRC) and other groups have attempted to assess progress in different areas of climate change science and to identify the critical research advances needed to further improve our understanding of past, current, and projected future climate changes; the impacts of these changes on both human and natural systems; and the infrastructure, organizational structures, and strategic frameworks needed to promote progress. Some of the most important sources consulted during the development of this paper are listed at the end.

This document was also informed by discussions held at CRC meetings during the past several years, including

- CRC Forum on Seamless Prediction and Year of Tropical Convection, May 17, 2007
- BASC/CRC Forum on IPCC AR4: Key Research Questions and High-Priority Research Needs, May 17, 2007
- CRC Forum on Integrated Earth System Analysis, March 22, 2006
- CRC Forum on Development of an Abrupt Climate Change Early Warning System, December 1, 2006

SELECTION CRITERIA

This paper attempts to distill the findings and recommendations from the reports and other sources listed in the preceding section down to a short list of the most important priorities for ensuring continued progress in understanding the natural climate system and climate change. We selected 15 priorities based on an informal, subjective consideration of the following criteria:

•   How important is the priority for documenting and understanding current climate change and its impacts?
•   How important is the priority for improving predictions and projections of climate variability and future changes in the climate system?
•   How relevant is the priority to decision support activities, including efforts to develop or evaluate strategies for responding to climate change?
•   What is the current level of support (within the CCSP, in the United States, and internationally) for progress on the priority, relative to its perceived scientific importance?
•   Can progress be made in the next 5 years given our current basis of understanding and available or potentially available technology and human resources?

We divided our 15 priorities into four categories: *overarching priorities*, which emerged as the most critically important for ensuring continued progress in climate change research; *existing priorities* from the CCSP's 2003 strategic plan, some of which have resulted in significant progress during the past 5 years while others have seen less; *emerging priorities*, or areas that have surfaced or increased in importance during the past 5 years; and *crosscutting priorities*, which involve the infrastructural and organizational frameworks needed to conduct climate research and connect the results to stakeholders. The following section provides a brief description of each priority. Our two overarching priorities are listed first and are viewed as top overall priorities for the CCSP. The remaining priorities are numbered for convenience but are not ranked, either with respect to one another or to priorities related to other aspects of the program. It should also be reiterated that these 15 priorities represent only a partial list of the full spectrum of valuable and worthwhile activities that could be undertaken to improve scientific progress on climate change, and are intended to serve as a starting point for continued discussion.

OVERARCHING PRIORITIES

1. *Observations*. Long-term, stable, and well-calibrated observations are fundamental to all climate research, prediction, and applications—as noted in the 2003 CCSP strategic plan (CCSP, 2003), "observations of the underlying physical state of the Earth system ... are required before questions about climate or global change can be addressed." Observations are likewise essential to improving our understanding and ability to model the individual components of the climate system, as well as how these elements interact. Hence, in our view, the single most important overall priority for the CCSP is the development of a sustained, integrated, and well-calibrated climate observing system that includes a broad spectrum of in situ and remotely sensed measurements from platforms in space, on land, at sea, and in the air. This includes maintaining the continuity of existing observations in order to determine anomalies and detect long-term changes, developing new observational capabilities targeting critical gaps, and integrating these elements to ensure and enhance the accuracy and comprehensiveness of the observing system.

The value of stable, homogeneous, long-term climate datasets has been recognized for many years (e.g., NRC, 1995, 1999, 2004a). Some examples of the research advances made possible by the current observing system can be found in the annual updates to *Our Changing Planet* (e.g., CCSP, 2007a), by comparing the most recent reports by the Intergovernmental Panel on Climate Change (IPCC, 2007) with previous assessments, and also in retrospective reports such as the recently released *Scientific Accomplishments of Earth Observations from Space* (NRC, 2008). Despite these important accomplishments, the use of existing observations for climate applications remains hampered by problems with precision, continuity, and spatial and temporal coverage, as well as a dearth of observations for key systems and an overall lack of coordination and integration. For example, despite repeated calls for investments in upgrading the surface observations network (e.g., NRC, 1999), climate trends from surface-based climate observations still contain significant errors (IPCC, 2007). More alarming, when the National Polar Orbiting Environmental Satellite System (NPOESS) experienced severe development problems and cost overruns, the climate research and monitoring instruments were demanifested; this and other concerns led the NRC Committee on Earth Sciences and Applications from Space to conclude that the current U.S. environmental satellite system is "at risk of collapse" (NRC, 2005a). Moreover, these ongoing and in some cases worsening problems have all occurred even though the 2003 CCSP strategic plan acknowledged the limitations of observing systems, dedicated an entire chapter to "Observing and Monitoring the Climate System," and called for the program to "expand observations, monitoring, and data/information system capabilities" (CCSP, 2003).

A number of factors have hampered the development of an integrated climate observing system. Climate observations demand dedicated long-term observational campaigns to evaluate climate variability of different timescales and estimate long-term trends. However, many "climate" observations are (or were) originally collected to support operational forecasting or other short-term applications that have less stringent accuracy and calibration requirements than climate monitoring and prediction. Maintaining long-term measurements also requires long-term agency attention and funding commitments, yet continuous climate observations unrelated to weather prediction do not appear to be the lead responsibility of any single agency. For example, for space-based platforms, the National Oceanic and Atmospheric Administration (NOAA) has traditionally performed operational data collection while the National Aeronautics and Space Administration (NASA) has conducted research data collection, but the relative role of each agency in maintaining records of suitable quality for climate research is less clear; this lack of clear responsibility may have led, in part, to the demanifestation of climate sensors from NPOESS and the Geostationary Operational Environmental Satellite-Series R (NRC, 2007f). Climate-relevant in situ data are managed by an even larger range of federal agencies, state agencies, international organizations, and other groups, creating additional obstacles to the development of an integrated climate observing system (e.g., NRC, 2007c). The Group on Earth Observations framework, a 10-year initiative launched in 2003, is intended to support and encourage international coordination in the development of an integrated observing system, but its progress to date and prospects for success given current funding levels remain unclear.

Without a coordinated, well-funded effort to design and build an integrated climate observing system, the climate record will remain fragmentary and inaccurate, compromising progress in all five CCSP goal areas. As the Climate Research Committee wrote in a letter to Senator Tim Wirth more than 10 years ago, "*Without this record, we cannot credibly assess natural climate variations, estimate anthropogenic effects on climate, judge the efficacy of negotiated mitigation efforts, or consider appropriate mid-course policy options*" (NRC, 1997). Observations are also critical to every single one of the suggested future priorities that follow; for instance, regional and mesoscale observing networks will be needed to support the development of regional climate models. Thus, in our view, observations should be made the top priority in the next CCSP strategic plan.

2. *Regional climate modeling*. Given a sound operational foundation in climate observations, a second overarching priority for U.S. climate science that we identified in many of the input documents listed in the "Context" section is improved regional-scale climate change predictions and projections. The regional level is where climate change, land use, and human pollutants intersect, and where climate change science connects most directly with decision makers and other stakeholders (e.g., NRC, 2003b). Our ability to model climate change at these scales has simply not kept pace with the increasing demand for this information. In our view, the lack of a national strategy to develop regional climate modeling capabilities is a fundamental shortcoming of the CCSP, and it means that the program has, to date, missed a major opportunity to connect climate change science to the citizenry.

Climate change projections from the current generation of global climate models are only reliable on continental-to-global scales, mainly because these models do a poor job of resolving smaller-scale climate variability. Although there have been a number of ad hoc efforts to project regional climate change by a variety of groups (e.g., Hay et al., 2002; NCAR, 2005), along with continuing improvements in the resolution of global models, these efforts have been limited by available computing resources (see priority 14, below) and by the lack of a well-developed, agreed-upon framework for regional downscaling, among other factors. There are also a number of important scientific and technical questions to be answered, for example:

- What are the potential benefits and drawbacks associated with statistical downscaling, nested models, stretched-grid global climate models, and other regional modeling approaches?
- What are the computational, observational, data assimilation, and other demands associated with each approach?
- What challenges can be expected when going to ultra-high resolution as a means of dealing with subgridscale parameterizations?

Given the importance of local-to-regional climate projections to decision makers, the CCSP should, in our view, make improved local and regional climate change prediction and projection a top priority. As with our first overarching priority on observations, regional modeling is a crosscutting effort that would facilitate progress in many of the CCSP goal areas, and it should be noted that these two overarching priorities are complementary and interdependent (e.g., observations are needed for model initialization, assimilation, forcing, and validation). The absence of adequate emphasis on regional climate change was also a concern expressed in the reviews of the current CCSP strategic plan (NRC, 2003b, 2004b), as was the coordination of modeling and observations at both regional and larger scales. Thus, we think it is imperative that the next CCSP strategic plan include regional climate modeling as a top priority, and that it contain a viable, integrated strategy for improving both these models and the observations on which they depend to ensure that local and regional decision makers have access to the information they need to plan for and respond to climate change.

EXISTING PRIORITIES

3. *Atmospheric distributions and effects of aerosols.* The 2003 CCSP strategic plan listed "advance[ing] the understanding of the distribution of all major types of aerosols and their variability through time, the different contributions of aerosols from human activities, and the processes by which the different contributions are linked to global distributions of aerosols" as one of the top three priorities for the Climate Change Research Initiative (CCRI), which was launched in 2001 to "leverage existing U.S. Global Change Research Program (USGCRP) research to address major gaps in understanding climate change." Over the past 5 years, significant progress has been made in measuring and characterizing the abundances and radiative impacts of certain kinds of aerosols, for example, the effects of black soot on snow and ice surfaces. However, the radiative forcing associated with aerosols, especially the so-called indirect effect of aerosols on cloud radiative properties, remains the single largest uncertainty in the total global radiative forcing associated with human activities (IPCC, 2007). Aerosols also have an important and complex influence on regional climate forcing, the implications of which are only beginning to be appreciated (NRC, 2005b).

In light of these remaining uncertainties and the rate of progress made to date, additional investments in the measurement, characterization, and modeling of aerosols could be expected to yield additional progress in understanding how aerosols impact both global and local climate. Encouragingly, the most recent update to *Our Changing Planet* (CCSP, 2007a) lists "understanding aerosol radiative forcing and interactions with clouds" as the first of eight interagency implementation priorities for FY 2008; the key objectives of this priority are "to quantify the effects of atmospheric aerosols (tiny airborne particles) on radiation and on clouds, to quantify the modification of the radiation balance by non-CO_2 greenhouse gases, and to quantify the influence of the chemistry of the lower atmosphere on both aerosols and non-CO_2 greenhouse gases." All of these objectives are, in our view, important and worthwhile. However, as with all of the other priorities listed below, continued progress in understanding aerosols depends on the two overarching priorities above, and will also require both focused and deliberate implementation planning and sustained funding commitments.

4. ***Climate feedbacks and sensitivity***. Climate feedbacks, initially focusing on polar feedbacks, and the overall sensitivity of the climate system were the second research priority identified for the CCRI (CCSP, 2003). As with the aerosol priority, considerable progress has been made over the past 5 years in improving understanding of several key individual feedback processes, including those associated with sea ice and water vapor (e.g., IPCC, 2007). However, the scope of climate processes and research activities that could be considered to fall into the general category of "climate feedbacks and sensitivity" is so broad that it is not clear that this actually constitutes a priority (e.g., many if not most of the activities that fall under CCSP goals 1 and 3 could be considered to fall under this heading). In addition, recent work (Roe and Baker, 2007) suggests that it will remain difficult to rule out the upper end of the probability distribution for the overall sensitivity of the climate system—that is, the possibility that Earth may warm much faster than the current midrange projections for a given future emissions scenario—even with continued progress in understanding climate feedbacks.

In our view, the CCSP would be better served to focus on a few key feedback processes that have both global and regional importance and that may be amenable to rapid progress. For example, in addition to the interaction between clouds and aerosols discussed in priority 3, above, there is also considerable uncertainty associated with how clouds will respond to rising temperatures (e.g., NRC, 2003c). Current climate models predict, rather than diagnose, cloud properties, and these predictions are now being tested against new data on cloud water content and other new global cloud observations. This should, over time, lead to greater realism in global cloud simulations in climate models. Other important feedback processes that may be amenable to progress include ice sheets and the carbon cycle, both of which are discussed below.

5. ***Carbon sources, sinks, and feedbacks***. Improving understanding of carbon sources and sinks, with a focus on North America, was the third "science" priority included in the CCRI. The U.S. Carbon Cycle Science Program, which is specifically

tasked with making progress in this area, has made significant progress in "clarifying the changes, magnitudes, and distributions of carbon sources and sinks; the fluxes between the major terrestrial, oceanic, and atmospheric carbon reservoirs; and the underlying mechanisms involved including human activities, fossil-fuel emissions, land use, and climate" (CCSP, 2007a). Many climate models are also now starting to include an explicit carbon cycle, improving the realism of climate change projections. Further details of progress in this area can be found in CCSP Synthesis and Assessment Product 2.2 *The First State of the Carbon Cycle Report* (CCSP, 2007c).

This recent progress does not mean, however, that there are not still a number of important research questions to be answered regarding the carbon cycle. For example, it will be critical to improve our understanding of the net carbon balance of high-latitude ecosystems, such as tundra and permafrost, as the climate continues warming, since this may be a key feedback on future warming due to the release of methane to the atmosphere (ACIA, 2004). Continued work on developing global carbon cycle models, and improving the linkages between these and other components in Earth systems models, would also be expected to improve our understanding of the interactions between carbon, ecosystems, agriculture, and hydrology, in addition to allowing better estimates of overall climate sensitivity and improved realism in future climate change projections. Thus, in our view, the CCSP should continue to include progress in this area as a program priority.

6. *Synthesis and assessment products*. The CCSP strategic plan (2003) called for the creation of a series of synthesis and assessment products that were intended to respond to "the top-priority research, observation, and decision support needs." While these documents have provided useful summaries of research progress and gaps in several important areas, to date only 3 of the 21 planned products have been released, and there have been concerns raised about the coordination of these activities, the significant time and resources needed to complete each assessment (especially given the time and resources recently dedicated to complete the IPCC assessment process), and the relevance of the synthesis and assessment products to decision makers, including those involved in the CCSP's own planning and budgeting process (e.g., NRC,

2003b, 2007a, d). In addition, the courts have ruled (Center for Biological Diversity v. Brennan, Court Order No. C 06-7062 BNA N.D. Cal. August 21, 2007) that the collection of 21 Synthesis and Assessment products does not fulfill the requirement of the 1990 Global Change Research Act for periodic national assessments of climate impacts. Hence, the emphasis on individual, topical synthesis and assessment products may be a priority that could be deemphasized or eliminated in the next CCSP strategic plan.

In our view, a true national assessment of current physical changes in climate and the impact of these changes on ecosystems and human systems, at both the regional and national levels, would also be much more valuable than a series of isolated reports on different topics. Without such an assessment, it is difficult for decision makers to evaluate the vulnerability of any particular region or sector to climate change, an important first step in improving resiliency to climate variability and change; when compounded by the lack of reliable regional model projections of future climate change (see priority 2, above), the dearth of knowledge about current and future climate change impacts on different sectors and regions also makes it difficult for decision makers to develop effective plans for responding to climate change through adaptation planning, mitigation efforts, and other activities. However, before a national assessment can be conducted, the United States first needs to develop a comprehensive, inclusive, well-thought-out, and carefully vetted strategy to produce it. Developing such a strategy should, in our view, be a top CCSP priority.

EMERGING PRIORITIES

7. *Integrated Earth system analysis (IESA)*. Our first emerging priority (i.e., area that has surfaced or increased in importance during the past 5 years) for the CCSP is IESA. Analysis (or reanalysis, when done retrospectively) refers to the synthesis of numerous, disparate observations, typically using a model, to provide a comprehensive, continuous, and physically consistent quantitative depiction of the temporal evolution of a climate system component; one notable example is the National Centers for Environmental Prediction (NCEP)/National Center for Atmospheric

Research atmospheric reanalysis (Kalnay et al., 1996). There has been good progress over the past decade in reanalyzing several individual system components, including the atmosphere and oceans (e.g., IPCC, 2007). Progress has been somewhat slower for other system components such as the cryosphere, land surface, and ecosystems, and efforts to link these separate analyses together are still in their infancy. In our view, an emerging priority for the CCSP should be the development of an IESA capability that assimilates the full range of Earth system observations, with the objective of creating an accurate, internally consistent, synthesized description of the evolving Earth system that is greater than the sum of its parts.

Since analysis is the bridge between observations and modeling, IESA would build on and expand our two overarching priorities. Many potential benefits of IESA have been identified, including the support of practical applications (e.g., agriculture, energy, and other economic sectors) and contributions to the Global Earth Observing System of Systems (GEOSS). IESA would also help address a previous criticism of the CCSP strategic plan (NRC, 2003b), namely, that observations and modeling efforts should be tied together more closely, for example, by using the observational record to critically assess model performance and using model simulations to assess the quality and adequacy of observations. The scientific and technical challenges associated with building an IESA capability include identifying the criteria for optimizing assimilation techniques for different purposes, estimating uncertainties, and meeting user demands for higher spatial resolution in analysis products.

8. *Ice sheets*. A number of recent studies have suggested that the Greenland and West Antarctic ice sheets may be less stable than previously thought, raising the possibility that global sea level rise could accelerate. For example, the downward percolation of surface meltwater can lubricate ice sheet movement, speeding ice flow, and destabilize floating ice shelves, as illustrated by the dramatic breakup of the Larson B ice shelf in 2002. The Larson B collapse also demonstrated that the removal of ice shelves or ice sheets can accelerate ice loss from surrounding glaciers, a process also now observed in Greenland. In addition, the Greenland ice sheet may be more vulnerable than previously thought to seawater

invasion at its margins. All of these concerns raise the possibility that ice sheets will discharge their ice volume to the sea more quickly, leading to more rapid sea level rise than the 1 to 2 feet projected for the late twenty-first century by IPCC (2007). There are signs that both the Antarctic and especially the Greenland ice sheet are already experiencing accelerating mass loss, but the error bars are wide.

Despite these recent advances in understanding, the response of ice sheets to global warming remains one of the largest uncertainties in projections of future climate change (IPCC, 2007). Ice sheet dynamics are in general poorly understood and poorly observed, and the dynamical response of ice sheets is completely unresolved in the current generation of climate models, and so the vulnerability of the Greenland and West Antarctic ice sheets to accelerated melting or collapse is thus poorly constrained. These gaps in understanding make it extremely difficult to place an upper bound on sea level rise during the twenty-first century and beyond. Rapid ice sheet disintegration, and the accompanying sharp increase in sea level rise, is also one of the most alarming potential mechanisms for future abrupt climate change, since even small changes in sea level are expected to have significant impacts on coastal communities and ecosystems around the globe. In our view, the CCSP should make the development of an ice sheet modeling capability in U.S. climate models a priority, including both the provision of computation resources and a plan to draw talent into the field, and should also continue to support ice sheet observations (including those from space; e.g., see NRC, 2007b) and monitoring during and after the International Polar Year.

9. *Decadal variability and abrupt climate change.* Our next suggested priority is improving our understanding of climate variability on decadal timescales and the nature and likelihood of possible abrupt climate changes during the twenty-first century. The problem of decadal climate variability is a major obstacle to understanding and predicting the El Niño/Southern Oscillation (ENSO) and also a major obstacle to understanding and predicting the regional effects of climate change; however, sorting out decadal variability from forced trends plus feedbacks is a major challenge. NRC (1998) identified the importance of the problem and made a single recommendation—that a national program in

decadal variability should be established. Unfortunately, such a program has not been established; although some progress has been made through CLIVAR and other domestic and international efforts, progress on this topic has generally been somewhat slow.

Understanding abrupt climate change is an important and closely related topic. Changes in the climate system are considered abrupt if they occur more rapidly than the time needed by ecosystems and society to adapt to them (NRC, 2002). In addition to rapid ice sheet collapse, described in the priority 8, other major mechanisms for abrupt climate change may include changes in radiative forcing (e.g., methane or carbon dioxide feedbacks), sea ice, the meridional ocean circulation, precipitation regimes (drought), and/or atmospheric circulation regimes (including tropical-extratropical feedbacks). Possible impacts include rapid sea level rise, severe and sustained droughts, or systematic changes in weather patterns over broad regions that may result from changes in ocean circulation (CCSP, 2007b). In our view, an important priority for the CCSP is to reduce the remaining knowledge gaps surrounding both decadal variability and abrupt climate change, and obtaining a better understanding of decadal/abrupt processes should be a prerequisite to attempting to develop an "early warning system" for detecting abrupt climate change.

10. *Nonstationary climate variability and seasonal-to-interannual prediction*. Currently, efforts to forecast how the climate may vary based on current conditions (climate predictions) and efforts to model long-term climate changes induced by changes in the natural and anthropogenic forcings (climate projections) are performed on separate research and organizational tracks in the United States. This separation impedes progress on critical questions at the intersection of climate prediction and projection, such as how the major modes of climate variability (e.g., ENSO, Pacific Decadal Oscillation, North Atlantic Oscillation, monsoons) will respond to global warming, and what the implications of these changes are for the skill of seasonal-to-interannual climate forecasts. Such questions are especially important in the context of our second overarching priority above (regional climate modeling), since one of the primary agents of regional climate change is these same regionally focused modes of climate variability. As a result, we lack a cohesive, integrated understanding of climate variability and predictability in the context of changing climate (IPCC, 2007).

The NRC *Review of the Final U.S. Climate Change Science Program Strategic Plan* (NRC, 2004b) states that the plan "presents a strategy for producing climate change projections through two modeling centers, but fails to present a national strategy for the seasonal to interannual climate predictions so important to many stakeholders" and that "without a fundamental change in approach to fully support seasonal to interannual climate prediction, the United States will be unsuccessful in the delivery of climate services." In our view, it is imperative that the next CCSP strategic plan include a strategy for moving toward a seamless and integrated suite of forecasting tools that spans the full range of timescales and modes of variability needed to make accurate predictions and projections. A necessary subelement of the above is a critical assessment of the present predictive skill of seasonal-to-interannual climate forecasts, as well as an evaluation of the potential predictability of the climate system on these and longer timescales. Without such an assessment, it will be difficult to develop a national strategy that integrates basic research, applied research, and application of climate forecasts on regional scales in direct support of the delivery of climate services (see also priority 14 below).

11. *Earth system predictability*. A number of federal agencies, departments, and programs produce forecasts and other environmental predictions. For instance, NOAA's NCEP delivers climate, weather, and ocean prediction products to a range of users, and is at the leading edge of efforts to improve predictive capabilities for these and other elements of the physical climate system. However, as distinct from 100-year projections, there does not appear to be a national strategy to expand, integrate, and improve these capabilities into a comprehensive environmental prediction system that combines weather forecasts, short-term climate predictions, and other components of the Earth system such as air quality, water quality, and terrestrial and ocean ecosystems. Ideally, such a strategy would ultimately lead to a robust operational infrastructure for an "end-to-end" process that accesses a huge array of global observations; assimilates them into interactive and coupled global atmospheric, land, and ocean models; runs weather and climate forecast models; and delivers forecast products to a diverse array of users.

The European Centre for Medium Range Weather Forecasting is actively moving toward a program for Earth system prediction that includes assimilation of the global carbon cycle, prediction of infectious disease outbreaks such as malaria, and seasonal forecasts for a range of agricultural crops. Similarly, in Japan the Earth Simulator supercomputer has served as a national focus for the development of a comprehensive Earth system model. In our view, the United States should also make Earth system prediction a priority. NOAA, NASA, and other agencies have already demonstrated international leadership with concepts such as GEOSS. A key to the success of GEOSS in the long term will be the sustained use and demand for Earth system observations in support of operational prediction across a broad range of sectors and Earth system components. This priority also has close connections to several of the other priorities suggested in this document, including the two overarching priorities on observations and regional modeling and the emerging priority on nonstationary prediction. Furthermore, the development of a predictive capability for the Earth system has unique policy relevance at both the national and international levels with respect to agriculture, ocean resources, energy, transportation, commerce, health, and homeland security. However, important questions will need to be answered before true Earth system predictions will be possible, such as the optimal pathway for developing predictive coupled physical–biological–chemical models of the Earth system, and the limits of predictability for different system components.

12. *Geoengineering*. Scientists and policymakers are beginning to appreciate that responding to climate change will require a portfolio of different responses (e.g., Pacala and Socolow, 2004). One possible, but controversial, response is geoengineering, or direct human intervention in the climate system intended to offset some aspects of climate change (NRC, 1992). Before policy makers can decide if geoengineering should play a role along with adaptation and mitigation efforts (for instance, if global warming occurs even more rapidly than the high end of the IPCC scenarios), a major research effort is needed to understand the efficacy, costs, and potential consequences and risks of the various geoengineering strategies that have been proposed, and to identify other potential alternative strategies. While there is a danger that some may inter-

pret geoengineering research as a "quick fix" to the climate problem that obviates critical adaptation and mitigation efforts, a failure to conduct careful research into different alternatives would be an even bigger risk. Articles on geoengineering by well-respected researchers are beginning to appear in the literature, but a more extensive research program in this area is needed. Such a program would need to involve both the CCSP and the Climate Change Technology Program, but the responsibilities, cost, deliverables, and form that such a program should take still need to be determined.

CROSSCUTTING (INSTITUTIONAL AND ORGANIZATIONAL) PRIORITIES[1]

13. *Computing (and storage)*. In our view, most if not all the priorities listed above are dependent on, and have to date been limited by, inadequate computational resources and the associated technical personnel. For example, regional climate modeling, climate reanalyses, and Earth system modeling have all been constrained by the lack of access to petascale computing, despite the current CCSP strategic plan (CCSP, 2003) listing "development of state-of-the-art climate modeling" as a CCSP priority. Storage issues have also gained urgency in light of the data management challenges associated with archiving and providing access to large quantities of data from new satellite missions and other high-volume observational streams (NRC, 2007c). In addition, the United States appears to lack a comprehensive strategy to marshal the considerable technical talent currently available in the private sector to address computing issues in climate research. Thus, in our view, a critical crosscutting priority for the CCSP is a realistic assessment of current and future computational requirements followed by the development of a comprehensive strategy for providing the requisite computational resources to support program activities. First and foremost among these is the issue of CPU speed and availability of overall computational horsepower/cycles,

---

[1] Our two overarching priorities (observations and regional modeling), could also be considered crosscutting priorities; however, we consider them sufficiently important to warrant elevating them to a higher level.

but other important issues include dealing with heterogeneous platforms, distributed/grid computing, and data archiving. Interagency coordination across the National Science Foundation, NASA, NOAA, and the Department of Energy and related engagement of the computer science and information technology communities will be essential to success.

14. ***Climate services***. As a result of the progress made to date within the CCSP and its predecessor USGCRP, the nation is poised to benefit in a routine manner from the transition from basic research to applied research to the provision of climate services—a mechanism to connect climate science to decision-relevant questions and support building capacity to anticipate, plan for, and adapt to climate fluctuations (Miles et al., 2006). Climate services, by definition, are "mission-oriented and driven by societal needs to enhance economic vitality, maintain and improve environmental quality, limit and decrease threats to life and property, and strengthen fundamental understanding of the earth" (NRC, 2001). While climate services may have some aspects in common with the mission and products of the National Weather Service, climate products are far more diverse and have unique requirements that go far beyond those associated with the day-to-day prediction of atmospheric conditions (e.g., Visbeck, 2008). However, there is currently not a single lead agency with the mandate and resources to operationally deliver climate services in response to stakeholder needs. What is needed, in our view, is a National Climate Service with many of the elements envisioned by Miles et al. (2006), for instance, organization at the federal level but taking advantage of the substantial regional-level expertise and experience (for instance, through the Regional Integrated Sciences and Assessments) needed to connect scientific results to individual stakeholders. Also, as called for in NRC (2001), the research enterprise dealing with environmental change and environment–society interactions should be enhanced in order to address the consequences of climate change and better serve the nation's decision makers, including "support of (a) interdisciplinary research that couples physical, chemical, biological, and human systems; (b) improved capability to integrate scientific knowledge, including its uncertainty, into effective decision support systems; and (c) an ability to conduct research at the regional or sectoral level that promotes

analysis of the response of human and natural systems to multiple stresses."

15. ***Integrated assessment***. Integrated assessment, our final crosscutting priority area, refers to the integrated analysis and modeling of the human activities and natural processes that give rise to greenhouse gas emissions and other climate forcings, the changes in the climate system caused by these forcings, the vulnerability and adaptive capacity of both human systems and natural systems to these changes in climate, and the estimated costs, benefits, and limitations of various mitigation and adaptation measures (e.g., Parson and Fisher-Vanden, 1995). By developing comprehensive models to address these issues, we could begin to provide decision makers with the information needed to make climate policy decisions that are both environmentally effective and economically efficient. The U.S. research community has a great deal of experience in the area of environmental assessments (e.g., NRC, 2007a), but integrated assessment represents a truly crosscutting activity that straddles even the traditional boundary between the natural and social sciences, which poses additional challenges and barriers. In our view, the CCSP should become more involved in leading and coordinating integrated assessment efforts—which should not be one-time events but a process that balances the needs of policy makers and the flow of information—to ensure that decision makers have access to the full range of information they need to respond to the many challenges of climate change.

REFERENCES

ACIA (Arctic Climate Impact Assessment), 2004, *Impacts of a Warming Arctic*, Synthesis Report, Cambridge University Press, Cambridge, 139 pp.

CCSP (Climate Change Science Program), 2003, *Strategic Plan for the U.S. Climate Change Science Program*, Climate Change Science Program and Subcommittee on Global Change Research, Washington, D.C., 202 pp.

CCSP, 2007a, *Our Changing Planet: The U.S. Climate Change Science Program for Fiscal Year 2008*, Climate Change Science Program and Subcommittee on Global Change Research, Washington, D.C., 212 pp.

CCSP, 2007b, *Summary of Revised Research Plan*, December 27, 2007, 12 pp., available at *http://www.climatescience.gov/Library/stratplan2008/summary/*.

CCSP, 2007c, *The First State of the Carbon Cycle Report (SOCCR): The North American Carbon Budget and Implications*, A.W. King, L. Dilling, G.P. Zimmerman, D.M. Fairman, R.A. Houghton, G. Marland, A.Z. Rose, and T.J. Wilbanks, eds., Synthesis and Assessment Product 2.2, Climate Change Science Program and Subcommittee on Global Change Research, Asheville, NC, 242 pp.

Hay, L.E., M.P. Clark, R.L. Wilby, W.J. Gutowski, G.H. Leavesley, Z. Pan, R.W. Arritt, and E.S. Takle, 2002, Use of regional climate model output for hydrologic simulations, *Journal of Hydrometeorogy*, **3**, 571–590.

IPCC (Intergovernmental Panel on Climate Change), 2007, *Climate Change 2007: Synthesis Report*, Contribution of Working Groups I, II, and III to the Fourth Assessment Report of the Intergovernmental Panel on Climate Change, Core Writing Team, R.K. Pachauri and A. Reisinger, eds., Geneva, 104 pp.

Kalnay, E., M. Kanamitsu, R. Kistler, W. Collins, D. Deaven, L. Gandin, M. Iredell, S. Saha, G. White, J. Woollen, Y. Zhu, A. Leetmaa, B. Reynolds, M. Chelliah, W. Ebisuzaki, W. Higgins, J. Janowiak, K. Mo, C. Ropelewski, J. Wang, R. Jenne, and D. Joseph, 1996, The NCEP/NCAR 40-year reanalysis project, *Bulletin of the American Meteorological Society*, **77**, 437–471.

Miles, E.L, A.K. Snover, L.C. Whitely Binder, E.S. Sarachik, P.W. Mote, and N. Mantua, 2006, An approach to designing a national climate service, *Proceedings of the National Academy of Sciences*, **103**, 19,616–19,623.

NCAR (National Center for Atmospheric Research), 2005, *Mesoscale and Microscale Meteorology (MMM) Division Science Plan: Five Years and Beyond*, available at *http://www.mmm.ucar.edu/about_mmm/stratplan/index.php*.

NRC (National Research Council), 1992, *Policy Implications of Greenhouse Warming: Adaptation, Mitigation, and the Science Basis*, National Academy Press, Washington, D.C., 944 pp.

NRC, 1995, *A Review of the U.S. Global Change Research Program and NASA's Mission to Planet Earth/Earth Observing System*, National Academy Press, Washington, D.C., 96 pp.

NRC, 1997, Letter to OSTP and Department of State on Global Observations of Climate, National Research Council, Washington, D.C., 2 pp. plus attachments.

NRC, 1998, *Decade-to-Century-Scale Climate Variability and Change: A Science Strategy*, National Academy Press, Washington, D.C., 160 pp.

NRC, 1999, *Adequacy of Climate Observing Systems*, National Academy Press, Washington, D.C., 50 pp.

NRC, 2001, *A Climate Services Vision: First Steps Towards the Future*, National Academy Press, Washington, D.C., 96 pp.

NRC, 2002, *Abrupt Climate Change: Inevitable Surprises*, National Academy Press, Washington, D.C., 244 pp.

NRC, 2003a, *Estimating Climate Sensitivity: Report of a Workshop*, National Academies Press, Washington, D.C., 62 pp.

NRC, 2003b, *Planning Climate and Global Change Research: A Review of the Draft U.S. Climate Change Science Program Strategic Plan*, National Academies Press, Washington, D.C., 85 pp.

NRC, 2003c, *Understanding Climate Change Feedbacks*, National Academies Press, Washington, D.C., 166 pp.

NRC, 2004a, *Climate Data Records from Environmental Satellites*, National Academies Press, Washington, D.C., 136 pp.

NRC, 2004b, *Implementing Climate and Global Change Research: A Review of the Final U.S. Climate Change Science Program Strategic Plan*, National Academies Press, Washington, D.C., 96 pp.

NRC, 2005a, *Earth Sciences and Applications from Space: Urgent Needs and Opportunities to Serve the Nation*, National Academies Press, Washington, D.C., 45 pp.

NRC, 2005b, *Radiative Forcing of Climate Change: Expanding the Concept and Addressing Uncertainties*, National Academies Press, Washington, D.C., 224 pp.

NRC, 2007a, *Analysis of Global Change Assessments: Lessons Learned*, National Academies Press, Washington, D.C., 196 pp.

NRC, 2007b, *Earth Sciences and Applications from Space: National Imperatives for the Next Decade and Beyond*, National Academies Press, Washington, D.C., 456 pp.

NRC, 2007c, *Environmental Data Management at NOAA: Archiving, Stewardship, and Access*, National Academies Press, Washington, D.C., 130 pp.

NRC, 2007d, *Evaluating Progress of the U.S. Climate Change Science Program: Methods and Preliminary Results*, National Academies Press, Washington, D.C., 170 pp.

NRC, 2007e, *NOAA's Role in Space-Based Global Precipitation Estimation and Application*, National Academies Press, Washington, D.C., 142 pp.

NRC, 2007f, *Options to Ensure the Climate Record from the NPOESS and GOES-R Spacecraft: A Workshop Report*, National Academies Press, Washington, D.C., 88 pp.

NRC, 2008, *Earth Observations from Space: The First 50 Years of Scientific Achievements*, National Academies Press, Washington, D.C., 144 pp.

Pacala, S., and R. Socolow, 2004, Stabilization wedges: Solving the climate problem for the next 50 years with current technologies, *Science*, **305**, 968–972.

Parson, E.A., and K. Fisher-Vanden, 1995, *Searching for Integrated Assessment: A Preliminary Investigation of Methods, Models, and Projects in the Integrated Assessment of Global Climatic Change*, Consortium for International Earth Science Information Network, University Center, Michigan.

Roe, G.H., and M.B. Baker, 2007, Why is climate sensitivity so unpredictable? *Science*, **318**, 629–632.

Visbeck, M., 2008, From climate assessment to climate services, *Nature Geoscience*, **1**, 2–3.

Appendix F

Workshop Agendas and Participants

WORKSHOP I ON FUTURE PRIORITIES FOR THE U.S. CLIMATE CHANGE SCIENCE PROGRAM
National Academy of Sciences
2101 Constitution Avenue, N.W.
Washington, D.C.
October 15–17, 2007

Agenda

Monday, October 15, Lecture Room

8:30   Plenary Session I: Overview [Chair: V. Ramanathan]

Introduction
- Goals and scope of the workshop
- Findings of *Evaluating Progress of the U.S. Climate Change Science Program: Methods and Preliminary Results* (NRC, 2007)
- Overview of the session
V. Ramanathan, Scripps Institution of Oceanography

8:40   Linkages Between Climate Science and Applications
S. Schneider, Stanford University

9:10   Overview of the U.S. Climate Change Science Program
(CCSP)   *W. Brennan, NOAA*

9:40   Evolving the CCSP to Meet the Needs of the Energy Industry
J. Chaudhri, SEMPRA

10:10   Evolving the CCSP to Help End Users
E. Claussen, Pew Center

10:40   Break

11:00   Committee's Initial Thoughts on Program Evolution
• Criteria for setting priorities
C. Justice, University of Maryland

Discussion   *All*

12:00   Working Lunch

1:00   Plenary Session II: National Perspectives
[Chair: V. Ramanathan]

Climate Change Science
S. Hays, Office of Science and Technology Policy

1:30   Instructions to the Working Groups   *C. Justice*
• Define criteria for prioritization
• Identify the top three to five priorities

1:45   Working groups convene

Working Group 1: Priorities for Applications
[Cochairs: J. Carberry, du Pont, and S. Schneider]
Working Group 2: Impacts, Adaptation, and Mitigation
[Cochairs: M.C. Lemos, University of Michigan, and
J. Edmonds, Joint Global Change Research Institute]
Working Group 3: Stakeholder Identification and
Interaction

[Cochairs: J. Jones, California Department of Water
Resources, and G. Eads, CRA International]
Working Group 4: Decision Support and Communication
[Cochairs: R. Kasperson, Clark University, and J.
Winkler, Michigan State University]

4:45   Plenary Session III: Initial Reactions, Concerns, and
        Crosscutting Issues

        Working Group 1   *J. Carberry or S. Schneider*
        Working Group 2   *M.C. Lemos or J. Edmonds*
        Working Group 3   *J. Jones or G. Eads*
        Working Group 4   *R. Kasperson or J. Winkler*

5:15   Reception

6:15   Workshop Adjourns for the Day

Tuesday, October 16, Members Room

8:30   Overview of Plans for the Day   *V. Ramanathan*

8:35   Plenary Session III: Working Group Reports
        [Chair: V. Ramanathan]

        Working Group 1   *J. Carberry or S. Schneider*
        Working Group 2   *M.C. Lemos or J. Edmonds*
        Working Group 3   *J. Jones or G. Eads*
        Working Group 4   *R. Kasperson or J. Winkler*

        Discussion

9:30   Plenary Session IV: Major Gaps and Future Priorities
        [Chair: V. Ramanathan]

        Observations to Support Climate Research and
        Applications   *K. Trenberth, NCAR*

10:00   Social Science Climate Data
                M. Hanemann, University of California, Berkeley

10:30   Break

10:45   Moving from Global- to Regional- and Local-Scale Models
                                            R. Leung, PNL

11:05   National Assessment
                A. Janetos, Joint Global Change Research Institute

11:30   Operational Research and Development
                            R. Balstad, Columbia University

11:55   Questions for the Speakers   *All*

12:30   Working Lunch

1:30   Toward Climate Services   *C. Koblinsky, NOAA*

2:00   Instructions to the Working Groups   *C. Justice*

2:15   Working groups convene to discuss near-term priorities for
        the CCSP

        Working Group 1: Priorities for Applications and How to
        Get There
            [Cochairs: A. Patrinos, Synthetic Genomics, and M.
            Hanemann]
        Working Group 2: Impacts, Adaptation, and Mitigation
            [Cochairs: M.C. Lemos and J. Edmonds]
        Working Group 3: Climate Services
            [Cochairs: G. Salvucci, Boston University, and R.
            Anthes, NCAR]
        Working Group 4: Assessments
            [Cochairs: S. Trumbore, University of California, Irvine,
            and W. Easterling, Pennsylvania State University]

5:30   Workshop Adjourns for the Day

Wednesday, October 17, Lecture Room

8:30   Plenary Session V: Stakeholder Needs and Modes of Interaction [Chair: M.C. Lemos]

        Overview of the Session

8:40   Climate Research Needs for the Energy Sector
S. Tierney, The Analysis Group

9:00   Climate Research Needs for the Water Sector
A. Watkins, New Mexico State Engineer's Office

9:20   Climate Research Needs for Reducing Greenhouse Gases
G. Franco, California Energy Commission

9:40   Questions for the Speakers   *All*

10:20   Break

10:40   Congressional Staff Panel

        M. Stephens, House Subcommittee on Interior, Environment and Related Agencies
K. Cook, House Subcommittee on Energy and Water
D. Butler, House Subcommittee on Energy and Water
J. Black, Senate Committee on Energy and Natural Resources

        Discussion   *All*

12:00   Working Lunch

1:00   Unmet Climate Science Communications Needs
A. Revkin, New York Times

1:30   Plenary Session VI: Working Group Reports

        Working Group 1   *A. Patrinos or M. Hanemann*
        Working Group 2   *M.C. Lemos or J. Edmonds*

Working Group 3	*G. Salvucci or R. Anthes*
Working Group 4	*S. Trumbore or W. Easterling*
Discussion	*All*

3:15   Break

3:30   Synthesis of Workshop Results
       Next Steps   *C. Justice and M.C. Lemos*

4:30   Workshop Adjourns

Workshop I Participants

David Allen, Climate Change Science Program Office
Rick Anthes, National Center for Atmospheric Research
Grayson Badgley, Georgetown University
Roberta Balstad, Columbia University
Nancy Beller-Simms, National Oceanic and Atmospheric
     Administration
Michele Betsill, Colorado State University
Rona Birnbaum, Environmental Protection Agency
Jonathan Black, Senate Committee on Energy and Natural
     Resources
William Brennan, National Oceanic and Atmospheric
     Administration
Dixon Butler, House Subcommittee on Energy and Water
Hannah Campbell, National Oceanic and Atmospheric
     Administration
Robert Cantilli, Environmental Protection Agency
John Carberry, E.I. du Pont de Nemours & Company
Javade Chaudhri, Sempra Energy
Ralph Cicerone, National Academy of Sciences
Eileen Claussen, Pew Center on Global Climate Change
Emily Therese Cloyd, Climate Change Science Program Office
Debra Conner, Environmental Protection Agency
Kevin Cook, House Subcommittee on Energy and Water
Mary Jane Coombs, West Coast Regional Carbon Sequestration
     Partnership

Mark Crowell, Federal Emergency Management Agency
Robert Curran, Climate Change Science Program Office
Roger Dahlman, Department of Energy
Brigid DeCoursey, Department of Transportation
Robert Dickinson, Georgia Institute of Technology
Randall Dole, National Oceanic and Atmospheric Administration
Kirstin Dow, University of South Carolina
George Eads, CRA International
Hallie Eakin, University of California, Santa Barbara
William Easterling, Pennsylvania State University
Jae Edmonds, Joint Global Change Research Institute
Jared Entin, National Aeronautics and Space Administration
Josh Foster, National Oceanic and Atmospheric Administration
Guido Franco, California Energy Commission
Teresa Fryberger, National Aeronautics and Space Administration
Chris Funk, University of California, Santa Barbara
Mary Glackin, National Oceanic and Atmospheric Administration
Patrick Gonzalez, Nature Conservancy
Anne Grambsch, Environmental Protection Agency
Michael Hanemann, University of California, Berkeley
Michelle Hawkins, National Oceanic and Atmospheric
     Administration
Sharon Hays, Office of Science and Technology Policy
Eileen Hofmann, Old Dominion University
Nate Hultman, Georgetown University
Anthony Janetos, Joint Global Change Research Institute
Jeanine Jones, California Department of Water Resources
Christopher Justice, University of Maryland
Roger Kasperson, Clark University
Chet Koblinsky, National Oceanic and Atmospheric Administration
Charles Kolstad, University of California, Santa Barbara
Ian Kraucunas, The National Academies
Martha Krebs, California Energy Commission
Greg Larson, City of Santa Cruz
Shirley Laska, University of New Orleans
Fabien Laurier, Climate Change Science Program Office
Linda Lawson, Department of Transportation
Anthony Leiserowitz, Yale University
Maria Carmen Lemos, University of Michigan

Teresa Leonardo, U.S. Agency for International Development
Fred Lestina, Georgetown University
Ruby Leung, Pacific Northwest National Laboratory
Anne Linn, The National Academies
George Luber, Centers for Disease Control and Prevention
Loren Lutzenhiser, Portland State University
Mike MacCracken, Climate Institute
Paola Malanotte-Rizzoli, Massachusetts Institute of Technology
Chad McNutt, National Oceanic and Atmospheric Administration
Ryan Meyer, Arizona State University
John Miller, Princeton Hydro
Ellen Mosley-Thompson, Ohio State University
Philip Mote, State of Washington
John Neuberger, Kansas University Medical Center
Claudia Nierenberg, National Oceanic and Atmospheric
    Administration
Benjamin Orlove, University of California, Davis
Kathryn Parker, Environmental Protection Agency
Stuart Parker, Inside Washington Publishers
Aristides Patrinos, Synthetic Genomics, Inc.
Rick Piltz, Climate Science Watch
Anne Polansky, Climate Science Watch
Thomas Pulzon, Georgetown University
Veerabhadran Ramanathan, Scripps Institution of Oceanography
Andrew Revkin, *New York Times*
Richard Richels, EPRI
Timmons Roberts, College of William and Mary
Rick Rosen, National Oceanic and Atmospheric Administration
Guido Salvucci, Boston University
Jason Samenow, Environmental Protection Agency
Roberto Sánchez-Rodríguez, University of California, Riverside
Stephen Schneider, Stanford University
Peter Schultz, Climate Change Science Program Office
Deborah Shapley, freelance writer
Caitlin Simpson, National Oceanic and Atmospheric
    Administration
Grant Smith, Dewberry
Michael Stephens, House Subcommittee on Interior, Environment
    and Related Agencies

Pamela Stephens, National Science Foundation
Paul Stern, The National Academies
Greg Symmes, The National Academies
Susan Tierney, The Analysis Group
Eric Toman, National Oceanic and Atmospheric Administration
Kevin Trenberth, National Center for Atmospheric Research
Susan Trumbore, University of California, Irvine
Robert Vallario, Department of Energy
Anne Watkins, New Mexico State Engineer's Office
Gene Whitney, Office of Science and Technology Policy
Bruce Wielicki, NASA Langley Research Center
Tom Wilbanks, Oak Ridge National Laboratory
Julie Winkler, Michigan State University
T. Stephen Wittrig, BP

WORKSHOP II ON FUTURE PRIORITIES FOR THE U.S. CLIMATE CHANGE SCIENCE PROGRAM
Millennium Harvest House, Century Room
1345 28th Street, Boulder, CO
March 19–20, 2008

Agenda

Wednesday, March 19 (Chair: V. Ramanathan)

8:30   Plenary Session I. Overview and Context

       Goals and scope of the workshop
       Emerging issues facing the nation
             V. Ramanathan, Scripps Institution of Oceanography

8:50   Urgent and Emerging Issues in Climate Change
                            T. Killeen, UCAR

9:20   Discussion: Are the Current CCSP Goals Right for the
       Future?   *All*

10:00   Break

10:20   Plenary Session II. Emerging Priority Areas

       User-Driven Priorities (from October 2007 workshop and
       Task 1 report)   *C. Justice, University of Maryland*

10:45   Priorities for the Human Dimensions of Climate
            T. Wilbanks, Oak Ridge National Laboratory

11:10   Priorities for Natural Climate Science
            A. Busalacchi, University of Maryland

11:35   Long-Range Perspectives for Climate Research
                     C. Koblinsky, NOAA

12:00   Working Lunch

1:00   Charge to the Working Groups   *V. Ramanathan*

Working groups convene to discuss emerging priority areas in the context of climate change science

Working Group 1: Social Science Priorities
    [Cochairs: M. Lemos, University of Michigan, and M. Hanemann, University of California, Berkeley]
Working Group 2: Natural Science Priorities
    [Cochairs: G. Salvucci, Boston University, and L. Dilling, University of Colorado]
Working Group 3: Natural Science Priorities
    [Cochairs: R. Dickinson, Georgia Institute of Technology, and D. Lettenmaier, University of Washington]
Working Group 4: Criteria for Prioritization
    [Cochairs: E. Mosley-Thompson, Ohio State University, and R. Leung, PNL]

4:00   Plenary Session III. Working Group Reports

Working Group 1   *M. Lemos or M. Hanemann*
Working Group 2   *G. Salvucci or L. Dilling*
Working Group 3   *R. Dickinson or D. Lettenmaier*
Working Group 4   *E. Mosley-Thompson or R. Leung*

5:30   Workshop Adjourns for the Day

Thursday, March 20 (Chair: C. Justice)

8:30   Overview of Plans for the Day   *C. Justice*

8:45   Plenary Session IV. Proposed Approach for Setting Priorities   *S. Trumbore, University of California, Irvine*

Discussion   *All*

9:30   Charge to the Working Groups   *C. Justice*

Working groups convene to test and fill in the matrix

Working Group 1, Cochairs: M. Lemos and D. Lettenmaier
Working Group 2, Cochairs: G. Salvucci and M. Hanemann
Working Group 3, Cochairs: R. Dickinson and L. Dilling
Working Group 4, Cochairs: E. Mosley-Thompson and R.
   Leung

11:30 Plenary Session V. Working Group Reports

Working Group 1   *M. Lemos or D. Lettenmaier*
Working Group 2   *G. Salvucci or M. Hanemann*
Working Group 3   *R. Dickinson or L. Dilling*
Working Group 4   *E. Mosley-Thompson or R. Leung*

12:30 Working Lunch

1:30   Plenary Session VI. Provocateurs on Program
       Infrastructure and Balance

Adequacy of Existing Modeling Infrastructure to Support
New Priorities   *L. Mearns, NCAR*

Adequacy of Existing Observation Infrastructure to Support
New Priorities   *A. Busalacchi, University of Maryland*

Ways to Balance Across the Program   *J. Fellows, UCAR*

2:15   Charge to the Working Groups   *C. Justice*

Working groups convene to discuss infrastructure and program balance

Working Group 5: Infrastructure to Support the Science
Priorities
       [Cochairs: E. Hofmann, Old Dominion University,
       and I. Held, Geophysical Fluid Dynamics Laboratory]

Working Group 6: Infrastructure to Support the Science
Priorities
    [Cochairs: J. Jones, California Department of Water
    Resources, and R. Lukas, University of Hawaii]
Working Group 7: Infrastructure to Support the Science
Priorities
    [Cochairs: R. Kasperson, Clark University, and D.
    Schimel, National Ecological Observatory Network]
Working Group 8: Balancing Basic and User-Driven Science
    [Cochairs: S. Trumbore and M. Krebs, California Energy
    Commission]

3:45   Plenary Session VII. Working Group Reports

        Working Group 5   *E. Hofmann or I. Held*
        Working Group 6   *J. Jones or R. Lukas*
        Working Group 7   *R. Kasperson or D. Schimel*
        Working Group 8   *S. Trumbore or M. Krebs*

4:45   Synthesis of Workshop Results
        Next steps   *V. Ramanathan and C. Justice*

5:30   Workshop Adjourns for the Day

Friday, March 21

8:30   Goals and Scope of the Meeting   *V. Ramanathan*
        Plans for the day

8:40   Panel on How to Make an Interagency Coordinated Program
        Work

        Management of Organizations
                                    R. Waterman, Waterman Group

        Business Perspective (*via telecon*)   *J. Carberry, du Pont*

Research Institutions Perspective
  A. MacDonald, Earth System Research Laboratory

Executive Office Perspective   *P. Backlund, NCAR*

10:15 Break

10:30 Discussion (led by R. Waterman)   *All*

11:15 Workshop Adjourns

Workshop II Participants

David Allen, Climate Change Science Program Office
David Anderson, National Oceanic and Atmospheric
    Administration
Joseph Arvai, Michigan State University
Peter Backlund, National Center for Atmospheric Research
Bruce Barkstrom, NOAA National Climatic Data Center
Ana Barros, Duke University
Robert Bindschadler, NASA Goddard Space Flight Center
Antonio Busalacchi, University of Maryland
Nancy Cavallaro, U.S. Department of Agriculture
Emily Therese Cloyd, Climate Change Science Program Office
James Coakley, Oregon State University
Roger Dahlman, Department of Energy
Eric Davidson, Woods Hole Research Center
Scott Denning, Colorado State University
Clara Deser, National Center for Atmospheric Research
Robert Dickinson, Georgia Institute of Technology
Lisa Dilling, University of Colorado
Kirsten Dow, University of South Carolina
Jay Fein, National Science Foundation
Jack Fellows, University Corporation for Atmospheric Research
David Halpern, National Aeronautics and Space Administration
Michael Hanemann, University of California, Berkeley
Isaac Held, NOAA Geophysical Fluid Dynamics Laboratory
Martin Hoerling, National Oceanic and Atmospheric
    Administration

Eileen Hofmann, Old Dominion University
Greg Holland, National Center for Atmospheric Research
James Hurrell, National Center for Atmospheric Research
Henry Jacoby, Massachusetts Institute of Technology
Marco Janssen, Arizona State University
Jeanine Jones, California Department of Water Resources
Christopher Justice, University of Maryland
Roger Kasperson, Clark University
Timothy Killeen, University Corporation for Atmospheric
    Research
Chester Koblinsky, National Oceanic and Atmospheric
    Administration
Charles Kolstad, University of California, Santa Barbara
Ian Kraucunas, The National Academies
Martha Krebs, California Energy Commission
Maria Carmen Lemos, University of Michigan
Robert Lempert, RAND Corporation
Dennis Lettenmaier, University of Washington
Lai-Yung Leung, Pacific Northwest National Laboratory
Anne Linn, The National Academies
Jennifer Logan, Harvard University
Roger Lukas, University of Hawaii
Michael MacCracken, Climate Institute
Alexander MacDonald, NOAA Earth System Research Laboratory
Mark McCaffrey, University of Colorado, Boulder
Michael McGeehin, Centers for Disease Control and Prevention
Chad McNutt, National Oceanic and Atmospheric Administration
Linda Mearns, National Center for Atmospheric Research
Gerald Meehl, National Center for Atmospheric Research
P.C.D. Milly, U.S. Geological Survey
Ellen Mosley-Thompson, Ohio State University
Carolyn Olson, U.S. Department of Agriculture
Bette Otto-Bleisner, National Center for Atmospheric Research
Kenan Ozekin, Awwa Research Foundation
Aristides Patrinos, Synthetic Genomics, Inc.
Ezekiel Peters, University of Colorado Natural Hazards Center
Roger Pulwarty, National Oceanic and Atmospheric
    Administration
Veerabhadran Ramanathan, Scripps Institution of Oceanography

A.R. Ravishankara, National Oceanic and Atmospheric
    Administration
Bob Raynolds, Denver Museum of Nature & Science
Rich Richels, Electric Power Research Institute
Eugene Rosa, Washington State University
Richard Rosen, National Oceanic and Atmospheric Administration
Guido Salvucci, Boston University
David Schimel, National Ecological Observatory Network
Peter Schultz, Climate Change Science Program Office
Joel Schwartz, Harvard University
David Skole, Michigan State University
Kirk Smith, University of California, Berkeley
Konrad Steffen, Cooperative Institute for Research in
    Environmental Sciences
Paul Stern, The National Academies
Taro Takahashi, Lamont-Doherty Earth Observatory
Lonnie Thompson, Ohio State University
Kathleen Tierney, University of Colorado
Susan Trumbore, University of California, Irvine
Susan Turnquist, Awwa Research Foundation
Peter van Oevelen, Global Energy and Water Cycle Experiment
Robert Waterman, Waterman Group
Anthony Westerling, University of California, Merced
Thomas Wilbanks, Oak Ridge National Laboratory

Appendix G

Biographical Sketches of Committee Members

Veerabhadran Ramanathan, *chair*, is a distinguished professor of atmospheric and climate sciences at Scripps Institution of Oceanography, University of California, San Diego. He received his Ph.D. in planetary atmospheres from the State University of New York, Stony Brook. Dr. Ramanathan's research focuses on global climate dynamics, the greenhouse effect, air pollution, and climate mitigation. A codiscoverer of the widespread South Asian atmospheric brown clouds (ABCs) in the late 1990s, he has since examined the impacts of ABCs on regional climate, including decreasing rice harvests in India and heating of the atmosphere over Asia and thus contributing to the melting of Himalayan and Tibetan glaciers. He currently chairs the United Nations Environment Programme-sponsored Project ABC. Dr. Ramanathan has been part of the United Nations Intergovernmental Panel on Climate Change (IPCC) since its inception, and served as one of the lead editors in the 2007 Working Group I report. He is the recipient of many national and international awards, including the Carl-Gustaf Rossby medal from the American Meteorological Society (AMS), the Buys Ballot medal from the Dutch Academy of Sciences, the Volvo environment prize, and the Zayed International prize for environment. Dr. Ramanathan is a member of the American Philosophical Society, U.S. National Academy of Sciences, Pontifical Academy of Sciences, Academia Europea, Third World Academy of Sciences, and Royal Swedish Academy of Sciences.

Christopher O. Justice, *vice chair,* is director of research and a professor in the Department of Geography at the University of Maryland. He holds a Ph.D. in geography from Reading University (UK). Dr. Justice has research interests in global environmental change, land-use and land-cover change, remote sensing, satellite-based fire monitoring, and terrestrial observing systems. He is the project scientist for NASA's Land Cover and Land Use Change Program and the Fire Implementation Team Leader for the Global Observation of Forest Cover project. He is also responsible for developing the Moderate Resolution Imaging Spectroradiometer fire product and rapid response system, a decision-making tool for resource managers. Dr. Justice is a former member of the National Research Council's (NRC's) Committee on Earth Studies. He is a current member of the Integrated Global Observation of Land theme and is leading the Group on Earth Observation Task on Global Agricultural Monitoring.

John B. Carberry recently retired as Director of Environmental Technology for the DuPont Company. At DuPont, he was responsible for analysis and recommendations for technical programs and product development for DuPont based on environmental issues. He led that technology function to provide excellence in treatment and remediation while in a transition to excellence in waste prevention, product stewardship and sustainability. Mr. Carberry presently consults (Carberry EnviroTech) on product and process strategies for dealing with the environmental issues of energy, renewable energy, sustainability, and nanomaterials. He is also an adjunct professor at both Cornell University and the University of Delaware. Mr. Carberry is a founding member of the Green Power Market Development Group. He recently chaired the NRC Committee on the Destruction of the Non-Stockpile Chemical Weapons, and served on six previous committees. He holds a B.ChE. and an M.E. in Chemical Engineering from Cornell University, an M.B.A. from the University of Delaware, and is a Registered Professional Engineer (Chemical).

Robert E. Dickinson is a professor in the Jackson School of Geosciences of the University of Texas at Austin. He received his Ph.D. in meteorology from the Massachusetts Institute of Technology. Dr.

Dickinson's research interests are in climate modeling, global change research, natural and anthropogenic forcing of climate variations, and land–atmosphere interactions in large-scale models. Dr. Dickinson has received a number of awards for his work in these areas, including the American Geophysical Union's (AGU's) Roger Revelle Medal and the AMS Rossby Award, Jule G. Charney Award, and Meisinger Award. He has participated in a number of climate-related committees, including the Climate Variability and Predictability Programme, the International Global Carbon Project (of the International Geosphere-Biosphere Programme, International Human Dimensions Programme, and World Climate Research Programme), and the NRC Committee on the Science of Climate Change. Dr. Dickinson is past president of the AGU, and a member of the National Academy of Engineering and the National Academy of Sciences.

Eileen E. Hofmann is a professor in the Department of Ocean, Earth and Atmospheric Sciences and a member of the Center for Coastal Physical Oceanography at Old Dominion University. She received her Ph.D. in marine sciences and engineering from North Carolina State University in 1980. Her research interests are in physical-biological interactions in marine ecosystems, climate control of diseases of marine shellfish populations, descriptive physical oceanography, and mathematical modeling of marine ecosystems. Dr. Hofmann has worked in a variety of marine environments, the most recent being the continental shelf region off the western Antarctic Peninsula. She currently coordinates the Southern Ocean Global Ocean Ecosystem Dynamics (SO GLOBEC) synthesis and integration effort and is an ex-officio member of the U.S. and International GLOBEC science steering committees. Dr. Hofmann has served on a number of NRC committees concerned with oceanography and ecology, including the Ocean Studies Board, the Committee on Ecosystem Management for Sustainable Marine Fisheries, and the Ecology Panel. She also brings expertise in evaluating research progress, having recently served on the NRC Committee on Metrics for Global Change Research.

James W. Hurrell is a Senior Scientist in and former Director of the Climate and Global Dynamics Division at the National Center for Atmospheric Research (NCAR). Although most of his professional career has been at NCAR, he spent a year as a visiting scientist at the UK Hadley Centre for Climate Prediction and Research. Dr. Hurrell received his Ph.D. in atmospheric science from Purdue University. His research has centered on empirical and modeling studies and diagnostic analyses to better understand climate, climate variability, and climate change. Dr. Hurrell has served on many national and international science planning efforts and is currently cochair of the Scientific Steering Group of the World Climate Research Programme on Climate Variability and Predictability. He has been extensively involved in assessment activities of the IPCC and the U.S. Climate Change Science Program (CCSP). He serves on the Council of the AMS, and is an AMS Fellow and recipient of the Society's Clarence Leroy Meisinger Award.

Jeanine A. Jones is a principal engineer and interstate resources manager at the California Department of Water Resources. She received her M.S. in civil engineering from the California State University, Sacramento, and is a registered civil engineer in California and Nevada. Ms. Jones was responsible for preparation of the 1998 update of the California Water Plan, the 2000 Governor's Advisory Drought Planning Panel report, and the 2008 California Drought Update report. She also participated in negotiations for the 2003 Colorado River Quantification Settlement Agreement and related agreements with relevant states and local agencies, and currently participates in the ongoing Colorado River Basin States discussions and Border Governors' Conference Water Worktable. Her statewide planning and drought management responsibilities included actions to inform the public about California drought vulnerability and to mitigate its effects. Such actions require the collection and analysis of regional data on parameters of interest to the CCSP, including land use, water use, water supply, and surface- and groundwater hydrology. Ms. Jones has served on the Colorado River Board of California and on a variety of committees of the Western States Water Council. She was also a governor's liaison to the Western Water Policy Review Advisory Commission.

Roger E. Kasperson is a research professor and distinguished scientist at Clark University. While at Clark University, he was also executive director of the Stockholm Environment Institute from 2000 to 2004. He holds a Ph.D. in geography from the University of Chicago. He has written widely on issues connected with risk analysis and communication, global environmental change, and environmental policy. Dr. Kasperson has served as a consultant or adviser to federal agencies and private entities on energy and environmental issues. Notable committee appointments include the Potsdam Institute of Climate Change Research Science Advisory Board, the UK Tyndall Institute for Climate Change Scientific Advisory Committee, Environmental Protection Agency Advisory Board, NRC Committee on the Human Dimensions of Global Change, and jury for the Volvo Environment Prize. He has been honored for his hazards research by the Association of American Geographers and was made a fellow of the American Association for the Advancement of Science (AAAS) and the Society for Risk Analysis for his contributions to the field of risk analysis. He is a member of the National Academy of Sciences and the American Academy of Arts and Sciences.

Charles D. Kolstad is a professor of environmental economics and policy at the University of California, Santa Barbara, where he holds joint appointments in the Bren School of Environmental Science and Management and the Department of Economics. He received his Ph.D. from Stanford in 1982. Dr. Kolstad's research interests are in environmental and natural resource economics, with a focus on environmental regulation and valuation. He is actively engaged in the economics of climate change and has a longstanding interest in energy markets. He was a participant in the U.S.-EU High-Level Transatlantic Dialogue on Climate Change in 2005 and is a lead author in the most recent assessment of the IPCC. Dr. Kolstad has been a member of several NRC committees concerned with climate, energy, and measuring program performance, including the Committee for Review of the U.S. Climate Change Science Program Strategic Plan, the Committee on Building a Long-Term Environmental Quality Research and Development Program in the U.S. Department of Energy, and the Board on Energy and Environmental Systems.

Maria Carmen Lemos is an associate professor of natural resources and environment at the University of Michigan and a senior policy analyst at the Udall Center for Studies of Public Policy at the University of Arizona. From 2006 to 2007 she was a James Martin Fellow at the Environmental Change Institute at Oxford University. She holds a Ph.D. in political science from the Massachusetts Institute of Technology. Her research interests focus on the human dimensions of global climate change, especially concerning the use of technical and scientific knowledge in climate-related policy and adaptation in less developed countries, the impact of technocratic decision making on democracy and equity, and natural resources (especially water) governance. Dr. Lemos has contributed to a number of national and international efforts related to climate change, including the IPCC Fourth Assessment (chapter on industry, settlement, and society) and CCSP syntheses and assessments on decision support experiments and evaluations using seasonal-to-interannual forecasts and observational data. She is a member of the Inter-American Institute for Global Change Research scientific advisory committee.

Paola Malanotte-Rizzoli is professor of physical oceanography in the Department of Earth, Atmospheric, and Planetary Sciences at the Massachusetts Institute of Technology. She is also director of the Joint Program in Oceanography and Ocean Engineering between MIT and the Woods Hole Oceanographic Institution. Dr. Malanotte-Rizzoli received her first Ph.D. in theoretical physics from the University of Padua (Italy) and her second Ph.D. in physical oceanography from the Scripps Institution of Oceanography. Her research interests are in modeling ocean circulation in specific basins and coastal seas; constraining ocean models with observations; modeling the Black Sea ecosystem; and studying tropical–subtropical interactions in the tropical Atlantic with emphasis on coupled ocean–atmosphere modes of variability. She also has practical interests in mitigating the impact of sea level rise and has been consulting on the project to build tidal gates in the Venice lagoon since 1995. She is a former president of the International Association for the Physical Sciences of the Oceans, a former member of the University Corporation for Atmospheric

Research Board of Trustees, and a former member of the National Science Foundation Advisory Committee for the Geosciences. She is a current member of the NRC Panel on Climate Variability and Change. She is a fellow of the AGU and the AMS.

Ellen S. Mosley-Thompson is a professor of climatology and atmospheric science in the Department of Geography, and a senior research scientist at the Byrd Polar Research Center at Ohio State University. She holds a Ph.D. in atmospheric science (geography) from Ohio State. Her research focuses on paleoclimate reconstructions from chemical and physical properties preserved in ice cores collected from Antarctica, Greenland, China, Africa, and South America. Dr. Mosley-Thompson has served on a number of NRC committees concerned with climate and polar regions, including the Committee on Glaciology, the Polar Research Board, and the Board on Global Change. She is a fellow of AAAS and a member of that association's steering group for geology and geography.

Aristides A.N. Patrinos is president of Synthetic Genomics. He received a diploma in mechanical and electrical engineering from the National Technical University of Athens and a Ph.D. degree in mechanical and astronautical sciences from Northwestern University. Following a brief research career, he joined DOE in 1988, where he led the development of DOE's program in global environmental change. From 1995 to 2006, he was the associate director for biological and environmental research in DOE's Office of Science, where he oversaw research activities in the human and microbial genome, structural biology, nuclear medicine, and global environmental change. He also directed the DOE component of the U.S. Human Genome Project and was the DOE representative to the CCSP and the Climate Change Technology Program. Dr. Patrinos is the recipient of numerous awards and honorary degrees, including three presidential rank awards for meritorious and distinguished service and two secretary of energy gold medals. He is a fellow of the AAAS and the AMS, and a member of the American Society of Mechanical Engineers and the AGU.

Guido D. Salvucci is a professor and chair of the Department of Earth Sciences and a professor in the Department of Geography and Environment at Boston University. He received his Ph.D. in hydrology from the Massachusetts Institute of Technology. His research focuses on coupled atmospheric water and energy balance processes, vadose zone hydrology, stochastic hydrology, and estimation of evapotranspiration and the water budget at large spatial scales through remote sensing. Dr. Salvucci has been active on hydrology committees and workshops, including the Consortium of Universities for the Advancement of Hydrologic Science's Standing Committee on Hydrologic Science, the NRC Committee to Review the GAPP Science and Implementation Plan, the Science Steering Group for the NASA Water Cycle Initiative, and the NRC Workshop on Groundwater Fluxes Across Interfaces. He is currently an editor of the Journal of Hydrometeorology. He is a recipient of the AGU's James B. Macelwane Medal and is also a fellow of that society.

Susan E. Trumbore is a professor in the Department of Earth System Science and at the University of California, Irvine. She received her Ph.D. in geochemistry from Columbia University. Her research interests are in the application of isotopes and tracers to problems in ecology, soil biogeochemistry, and terrestrial carbon cycling. Dr. Trumbore was an author of the IPCC report on land use, land-use change, and forestry. In addition to her teaching and scientific pursuits, she is interested in the evaluation of research programs and served on the NRC Committee on Metrics for Global Change Research. She is a fellow of the AAAS and president-elect of the AAAS Geography and Geology section. She is a fellow of the AGU and a former president of AGU's biogeochemistry section. She is a member of the Max Planck Society and will assume a directorship at the Max Planck Institute for Biogeochemistry, in Jena, Germany.

T. Stephen Wittrig is director of advanced technologies at BP. He received his Ph.D. in chemical engineering from the California Institute of Technology. Dr. Wittrig is responsible for BP's academic and external technology programs in Russia and China. His current work focuses on developing a long-term technology strategy for BP, emphasizing clean energy technologies (solar, wind, hydrogen, and combined-cycle gas-turbine power generation) and techniques for sequestering CO_2 in depleted oil reserves. In previous positions at Amoco, he helped develop strategies for converting gas to liquids and oxygenates and for implementing chemical technologies, managed the engineering and process evaluation group for new chemical products development, and led a team to develop new reactor technology for converting methane to syngas. Dr. Wittrig was a member of the NRC committee that reviewed the CCSP strategic plan in 2004.

Appendix H

Acronyms and Abbreviations

CCSP   Climate Change Science Program
CCTP   Climate Change Technology Program
CEOS   Committee on Earth Observation Satellites
CRA   Comparative Risk Assessment
DICE   Dynamic Integrated Climate Economy
DOE   Department of Energy
EIA   Energy Information Administration
ENSO   El Niño/Southern Oscillation
EOS   Earth Observing System
EPA   Environmental Protection Agency
ESA   Endangered Species Act
GCOS   Global Climate Observing System
GEOSS   Global Earth Observing System of Systems
GOOS   Global Ocean Observing System
GTOS   Global Terrestrial Observing System;
IAI   Inter-American Institute for Global Change Research
IGBP   International Geosphere-Biosphere Programme
IHDP   International Human Dimensions Programme on Global Environmental Change
IPCC   Intergovernmental Panel on Climate Change
NASA   National Aeronautics and Space Administration
NOAA   National Oceanic and Atmospheric Administration

NPOESS	National Polar-orbiting Operational Environmental Satellite System
NRC	National Research Council
NSF	National Science Foundation
PDSI	Palmer Drought Severity Index
RISA	Regional Integrated Sciences and Assessments
SAR	synthetic aperture radar
START	Global Change System for Analysis, Research, and Training
USDA	U.S. Department of Agriculture
USGCRP	U.S. Global Change Research Program
WCRP	World Climate Research Programme
WHO	World Health Organization